NATURAL ENVIRONMENT RESEARCH COUNCIL

INSTITUTE OF TERRESTRIAL ECOLOGY

# ECOLOGY OF QUARRIES

## The importance of natural vegetation

ITE SYMPOSIUM NO. 11

Edited by:

B N K DAVIS

ITE, Monks Wood Experimental Station,
Abbots Ripton, Huntingdon

Proceedings of a Workshop held at Monks Wood Experimental Station
23-24 February 1981

Printed in Great Britain by NERC/SERC Reprographic Services, Swindon



First published in 1982 by Institute of Terrestrial Ecology
68 Hills Road
Cambridge
CB2 1LA
0223 (Cambridge) 69745

ISBN 0 904282 59 7

ACKNOWLEDGEMENT

The figures were drawn by Miss Sarah Anthony and the plates done by P G Ainsworth. Camera-ready copy was prepared by Mrs Di Plant and Mrs Barbara Stocker.

COVER PHOTOGRAPH

Natural revegetation at Bowers quarry, Portland Bill, Dorset in 1975. Photograph B N K Davis.

*The Institute of Terrestrial Ecology (ITE)* was established in 1973, from the former Nature Conservancy's research stations and staff, joined later by the Institute of Tree Biology and the Culture Centre of Algae and Protozoa. ITE contributes to and draws upon the collective knowledge of the fourteen sister institutes which make up the *Natural Environment Research Council*, spanning all the environmental sciences.

The Institute studies the factors determining the structure, composition and processes of land and freshwater systems, and of individual plant and animal species. It is developing a sounder scientific basis for predicting and modelling environmental trends arising from natural or man-made change. The results of this research are available to those responsible for the protection, management and wise use of our natural resources.

One quarter of ITE's work is research commissioned by customers, such as the Nature Conservancy Council, who require information for wildlife conservation, the Department of Energy and the Department of the Environment, and the EEC. The remainder is fundamental research supported by NERC.

ITE's expertise is widely used by international organisations in overseas projects and programmes of research.

DR B N K DAVIS
Institute of Terrestrial Ecology
Monks Wood Experimental Station
Abbots Ripton
HUNTINGDON Cambridgeshire
PE17 2LS

04873 (Abbots Ripton) 381

# CONTENTS

# INTRODUCTION

**B N K DAVIS**
*Institute of Terrestrial Ecology, Monks Wood Experimental Station, Huntingdon.*

Many of the major problems of land reclamation in this country have been overcome by engineers, landscape designers and agriculturalists. Increasingly, however, we are faced with questions over the restoration or future role of mineral workings which cannot readily or entirely be returned to agriculture, forestry, industry or other economic use. This workshop was called to consider especially the ecological aspects of hard rock quarries. These quarries have attracted interest from ecologists in recent years, not only because of the natural processes of revegetation that they present for study, but because of the need to increase the range of options available to planners, landscape architects and conservationists. Land and vegetation are basic resources, and it is clear that we cannot rely entirely on nature's slow powers of healing in the future; natural processes cannot keep up with the rate of working or the rate of destruction of semi-natural habitats from which recolonisation could take place.

The workshop was divided into three sessions. The first three papers set the scene by examining the present state of quarries and pointing to the range of ecological and land-use questions that they pose. What environmental or biological factors control the natural development of vegetation in quarries? Why are some quarries important nature reserves now and how should they be managed? The subsequent sessions dealt with more specific questions. How do plants colonise man-made habitats? What determines plant establishment and succession, and what corresponding changes occur in the soil? How can we direct or modify quarrying techniques and revegetation to produce certain goals - and what goals are desirable? This symposium has answered some of these questions but its real value may be judged by the stimulus that it gives to the application of existing knowledge and to further study of the problems.

A discussion was held after each session but many of the earlier themes were discussed more fully at the end and so edited summaries of the main points are presented here in two parts.

# STATE AND STATUS

## THE BOTANICAL INTEREST AND VALUE OF QUARRIES

**J G HODGSON**
*Unit of Comparative Plant Ecology (NERC), Sheffield*

### INTRODUCTION

Quarrying is an economically essential land use that has affected the vegetation of Britain in a variety of ways. It has resulted in the destruction of irreplaceable relict plant communities (Ratcliffe 1974), and the deposition of limestone dust resulting from quarrying may also have a more deleterious effect on the surrounding vegetation than is generally appreciated (Manning 1971; Etherington 1978). However, some quarries are of considerable value for conservation (Ratcliffe 1974; Davis 1976; Holliday & Johnson 1979) and it is this positive aspect of quarries that will be considered by reference to the following questions:

1. Are quarries important because of their floristic composition and, if so why?

2. How relevant to reclamation procedures is an understanding of natural colonisation processes?

### QUARRY HABITATS

While they may show a wide spectrum of environmental conditions, quarries typically contain rocky habitats with or without areas of overburden. Their shallow soils are often very immature and frequently infertile because of such factors as nutrient deficiency, drought and erosion (Goodman 1974; Holliday & Johnson 1979; Johnson & Bradshaw 1979; Bradshaw & Chadwick 1980). Two types of quarry may be recognised - calcareous (associated with the chalk and limestone) and acidic (gritstone, slate, granite etc.). In the latter type of quarry, infertility is combined with the problems of soil acidity. From the preceding description, it is clear that the plants best adapted to quarry conditions are not the productive species of agricultural systems, but plants characteristic of infertile environments.

### THE VEGETATION TYPICALLY ASSOCIATED WITH QUARRIES

*Study area*

The Sheffield region rather than the whole of Britain is chosen for analysis because the results of three recent and extensive vegetation surveys are available (semi-natural grassland, Grime & Lloyd 1973; major habitats, Grime, J.P., Hodgson, J.G. & Hunt, R., in preparation; rare plant species and communities, Hodgson, J.G. *et al.*, in preparation). The region covers an area of 2330 $km^2$. It includes upland, acidic quarries in the Coal Measure and Millstone Grit Sandstone, and both upland (Carboniferous Limestone) and lowland (Magnesian Limestone) calcareous workings. Sand and gravel pits do not fall strictly within the dictionary definition of quarries - 'open excavations for building-stone, slate, etc.' - and will not therefore be considered. All uncited descriptions of the ecology of species in the ensuing text are derived from the three local surveys.

*Typical vegetation*

Data from the survey of major habitats (Table 1) indicate that the vegetation of calcareous spoil is moderately species-rich with more forbs than grasses, and with a significant component of legumes and annuals. Shrubs may or may not be present. On acidic spoil, there are fewer species and almost no legumes or annuals. However, heather (*Calluna vulgaris*) and other low-growing shrubs are often found.

**TABLE 1 The composition of vegetation typically associated with quarry spoil in the Sheffield region. (Unpublished data from Survey of Major Habitats.)**

| | Mean no. spp./$m^2$ | | | | | |
|---|---|---|---|---|---|---|
| | Total | Grasses | Forbs | Woody spp. | Legumes | Annuals |
| Calcareous | | | | | | |
| Upland (Carboniferous Limestone) | 14.5 | 4.1 | 10.3 | 0.1 | 0.8 | 2.3 |
| Lowland (Magnesian Limestone) | 18.9 | 4.4 | 13.2 | 1.3 | 0.9 | 4.3 |
| Acidic | | | | | | |
| Upland (Millstone Grit and Coal Measures) | 4.5 | 1.9 | 1.3 | 1.3 | 0.1 | <0.1 |

The most widespread quarry species are plants of grassland, heath and open habitats reflecting the range of environments most frequently encountered within quarries (Table 2).

The species listed for acidic spoil are all adapted to the acidic soils of their quarry environment and all are common components of the semi-natural vegetation usually adjacent to locally-occurring acidic quarries. It is not known whether the species are mobile, but the two grasses in particular are found in a wide range of man-made habitats.

Surprisingly the most consistent colonists of calcareous sites are not, as might have been predicted, species particularly adapted to a limestone environment. Nine (69% of the commonest species of quarry spoil) are amongst the twenty most widespread components of the local flora, and, with one exception, they are all plants associated with a wide range of vegetation types and soil conditions.

These quarry species are also readily divisible into a) very common grasses and b) common, but less widespread, forbs with a wind-dispersed fruit or seed. Forbs without this dispersal mechanism are less frequent. Thus, the most widespread species of calcareous quarry spoil lack specificity to calcareous habitats, may be very mobile, and are all common plants. This result suggests that the nature of the vegetation close to quarries and the mobility of its component species exert a critical effect on the pattern of colonisation, a theme that will be returned to later.

**TABLE 2 The commonest species (>50% occurrence) of typical quarry spoil. Data are also provided as to whether each species is amongst the fifty most commonly recorded species in the Survey of Major Habitats.**

| | % occurrence | Ranking within a list of the commonest species of Sheffield region |
|---|---|---|
| Carboniferous Limestone | | |
| Grasses | | |
| *Arrhenatherum elatius* (false oat-grass) | 68 | 9 |
| *Dactylis glomerata* (cocksfoot) | 50 | 7 |
| *Festuca rubra* (red fescue) | 84 | 3 |
| *Poa pratensis* (smooth meadow grass) | 55 | 8 |
| Forbs with wind-dispersed seed | | |
| *Hieracium* sect. *Hieracium* (hawkweed) | 63 | 19 |
| *Leontodon hispidus* (rough hawkbit) | 72 | >50 |
| *Tussilago farfara* (coltsfoot) | 50 | 39 |
| Magnesian Limestone | | |
| Grasses | | |
| *Agrostis stolonifera* (creeping bent) | 63 | 6 |
| *Dactylis glomerata* (cocksfoot) | 63 | 7 |
| *Holcus lanatus* (Yorkshire fog) | 67 | 5 |
| Forbs with wind-dispersed seed | | |
| *Chamaenerion angustifolium* (rosebay willow-herb) | 52 | 11 |
| *Crepis capillaris* (smooth hawk's-beard) | 56 | >50 |
| *Senecio jacobaea* (ragwort) | 63 | 33 |
| *Taraxacum* agg. (dandelion) | 56 | 13 |
| Millstone Grit and Coal Measures | | |
| Grasses | | |
| *Deschampsia flexuosa* (wavy hair-grass) | 92 | 2 |
| Other species | | |
| *Calluna vulgaris* (heather) | 57 | 46 |
| Less widespread species | | |
| *Agrostis tenuis* (common bent) | 35 | 4 |
| *Galium saxatile* (heath bedstraw) | 24 | 36 |

## THE VEGETATION OF QUARRIES WITH A CONSERVATION INTEREST

### *The significance of rare plants*

The presence and beauty of rare species are often used as an argument for conservation. Certainly, rare species provide an intriguing scientific problem because the reasons for the rarity of many of them are not understood. However, rare species are also perhaps the most specialised and fragile components of ancient semi-natural communities. Such vegetation is of great scientific importance as it provides an historical record of the combined effects of natural processes and land use. It is also a vital part of our natural heritage, a relic, along with ancient buildings, of the landscape in which our ancestors lived.

The relevance of rare plants, in assessing the conservation value of quarries, would be enhanced if it could be demonstrated that the vegetation associated with such sites has features in common with ancient plant communities. Spoil is one of the most important habitats within quarries. Accordingly, a simple comparison was made between the commonest species of this habitat associated with a) randomly sampled (typical) vegetation, b) vegetation containing rare species and c) the commonest plants of old semi-natural grassland.

The preliminary findings for limestone quarries (Table 3) suggest that the vegetation with rare plants is intermediate between that of typical spoil and that of ancient semi-natural grassland. One anomaly is the low value for species density in semi-natural grassland on the Magnesian Limestone. This result may be explained by the fact that only 12% of the sampled areas were still grazed. Data for open habitats within quarries, not presented here, show similar trends when compared with naturally-occurring rocky areas of floristic interest. Thus, rare plants may be used as indicator species for vegetation with at least some of the characteristices of prime, ancient plant communities.

### *An assessment of the conservation value of quarries*

If the habitats within quarries were very unusual, one would expect many species to be confined to quarries. Since only 7 out of the 270 rare species of the region occur predominantly within quarries, it appears that the quarry environment is not unique. Despite their small surface area within the landscape, quarries contain at least one site for 27% of all the rare native species recorded for the Sheffield region (Table 4). If the ecology of these species is examined, it is found that 21% of all rare plants are associated with calcareous and 7% with acidic quarries. Furthermore, quarries contain 56% of all rare species associated with grassland + heathland and 36% of the rare plants from open habitats (Table 5).

Open, infertile, naturally occurring rocky environments with small pockets of soil have been important refugia for many rare species since historical times (Pigott & Walters 1954). The occurrence of some of these species adds considerably to the floristic interest of quarries.

In addition, ancient vegetation has decreased catastrophically both in quantity and quality, particularly in lowland Britain (Duffey 1973; Ratcliffe 1974; Nature Conservancy Council 1981). There are, for example, now only five areas of grazed ancient grassland (two of which are old quarries) within the 250km$^2$ of the Magnesian Limestone of the Sheffield region. Three of these sites need to be preserved to retain what remains of

TABLE 3 A comparison of the vegetation of a) typical quarry spoil, b) quarry spoil with rare species and c) ancient grassland with particular reference to the characteristics of the commonest species of each habitat. (Unpublished data from Surveys of Semi-natural Grassland, Major Habitats and Rare Plant Species and Communities.)

With the exception of the species/metre$^2$ value, all the data presented refer to the attributes of the twenty commonest species of each category. "Mobile species" include only plants whose seeds have a plume of hairs and are transported by the wind. "Very common species" refer to the number of species within the twenty most frequently recorded in the Survey of Major Habitats. "Restricted grassland species" include the twenty species most commonly recorded in semi-natural grassland of the same geology that are not "very common species". Rankings for each set of attributes are included in parentheses as an aid to comparison.

| | Spp./m$^2$ | No. of mobile spp. | No. of very common spp. | No. of restricted grassland spp. | No. of species in common with ancient grassland |
|---|---|---|---|---|---|
| Carboniferous Limestone | | | | | |
| Typical | 14.5 (3) | 7 (1) | 11 (1) | 4 (3) | 9 (3) |
| + Rare species | 20.9 (2) | 3.75 (2) | 8 (2) | 5 (2) | 11 (2) |
| Ancient grassland | 22.3 (1) | 1 (3) | 8 (2) | 12 (1) | 20 (1) |
| Magnesian Limestone | | | | | |
| Typical | 18.9 (3) | 10 (1) | 11 (1) | 3 (3) | 6 (3) |
| + Rare species | 23.5 (1) | 3 (3) | 8 (2) | 10 (2) | 11 (2) |
| Ancient grassland | 20.6 (2) | 4 (2) | 5 (3) | 15 (1) | 20 (1) |

TABLE 4 The number of locally rare native species found in quarries in the Sheffield region. (Unpublished data from Survey of Rare Plant Species and Communities.)

(Very rare, scarce and uncommon species are confined to <5%, <10% and <20% of the 120 5-km squares of the survey area respectively.)

| Status | No. spp. | % |
|---|---|---|
| Very rare | 15 | 14 |
| Scarce | 28 | 34 |
| Uncommon | 30 | 41 |
| Total | 73 | 27 |

TABLE 5 The habitats with which the rare species of quarries are usually associated in the Sheffield region. (Unpublished data from Survey or Rare Plant Species and Communities).

| | Total no. of species | % rare species |
|---|---|---|
| Grassland + heathland | 29 | 56 |
| Open habitats | 20 | 36 |
| Woodland | 10 | 29 |
| Marsh | 13 | 18 |
| Arable | 1 | 8 |
| Aquatic | 0 | 0 |

the floristic diversity of grazed Magnesian Limestone grassland. It is in this context that the vegetation of quarries needs to be considered, particularly in lowland areas where pressures of land use are most intense. Thus, it is clear that quarries, particularly on calcareous substrates, are of considerable value. Their importance as refugia for rare plants and vegetation types akin to ancient grassland communities will increase further as more areas of this vegetation are destroyed for agriculture and industry, or are left unmanaged to revert to coarse grassland and scrub.

## THE ROLE OF COLONISATION IN DETERMINING THE CONSERVATION INTEREST OF QUARRIES

The slow rates of both soil formation and the revegetation of quarries are critical features of the process of colonisation. Many quarries on the Carboniferous Limestone of Derbyshire, for example, still contain large areas of open vegetation even after 50 years, and Welsh slate quarries of comparable age are even more poorly colonised, being virtually devoid of higher plants. Even when a more readily weathered rock such as chalk is involved, the processes of colonisation and succession are slow (Hope-Simpson 1940). Since most are ungrazed, the more mature quarries will eventually lose much of their floristic interest through the growth of tall grasses or scrub. However, many quarries have apparently always lacked floristic interest, and a possible reason for this deficiency will be considered by reference to 3 very different limestone quarries within the Sheffield area:

*1.*
The first example is a small, ancient, rather overgrown quarry on the Magnesian Limestone. It contains two-thirds of the species restricted to, or commonest in, calcareous quarries and also two locally rare species, one of which, *Carex ericetorum*, is nationally rare (David 1981). The site, now a golf course, was formerly an old common and is designated a Site of Special Scientific Interest. It is easy to envisage the colonisation in former times of this small quarry by native species from the adjoining vegetation as the surrounding area is infertile and calcareous. Ironically, now that the common is no longer grazed, the quarry, with its shallow soil, has by far the more species-rich vegetation.

*2.*
The second is Millers Dale Quarry on the Carboniferous Limestone within the Peak Park (see Holliday & Johnson 1979). It is much younger, being disused for only about 50 years and is surrounded by ancient, semi-natural vegetation which, together with the quarry, is designated a local nature reserve. The quarry is very rich floristically. A few of the species present, notably several orchids, have probably colonised from a distance. The rest are mostly found in the adjacent vegetation and invasion was presumably from

this source. Two new species are known to have become established within the last five years and many potential colonists are present in the surrounding vegetation. This process of colonisation of one set of infertile habitats by species of another is a major cause of the floristic interest of the quarry.

3.
The third example, an active quarry of no floristic interest on the Magnesian Limestone, is surrounded by fertile arable land. The nearest ancient vegetation is three miles away. Under these circumstances, colonisation from adjacent fertile habitats is probable and will result in the invasion of species only poorly adapted to the quarry environment. In time, quarry ecotypes might be evolved and the species diversity may be augmented by the invasion of very mobile species from a greater distance (Gemmell 1980) and by the introduction of plants by quarry traffic.

To summarize, it is suggested that the nature of nearby plant communities is a very important factor in colonisation and that it is the proximity during colonisation of ancient vegetation that has resulted in most of the floristic interest associated with quarries today.

The results in Table 3 indicate that the vegetation of typical quarry spoil on the Carboniferous Limestone is more like ancient grassland than is quarry vegetation on the Magnesian Limestone. They suggest that there are crucial differences between the Carboniferous Limestone and the Magnesian Limestone in the range of species available for the colonisation of quarries. This is because substantial areas of ancient Carboniferous Limestone grassland still occur within and outside the Peak District National Park, whereas only a few fragments of ancient Magnesian Limestone vegetation remain. Thus, further evidence is provided that the nature of the adjacent vegetation is very important in determining the nature of the colonisation.

If the processes of colonisation are considered in more detail, it would be predicted that limestone spoil at an early stage would a) be species-poor since colonisation is slow, b) contain a large proportion of species with an effective mechanism for long-distance dispersal, c) support many of the commonest species of the landscape and d) support only a small number of less frequent species. In contrast, ancient limestone grassland may be expected to show the opposite trends. Table 3 indicates that typical quarry spoil and ancient grassland do differ in these respects and that quarry vegetation containing rare species is intermediate and may represent a half-way successional state. Therefore, it is suggested that, with some exceptions (notably several orchids), rare plants are not primary colonists and their presence may be taken to indicate a quarry where both colonisation from adjacent semi-natural vegetation and succession has been occurring for a considerable time (see Usher 1973).

The data of Hope-Simpson (1940) suggest that species were still invading a chalk pit even after 40 years and are in agreement with the observations for Millers Dale Quarry given here. Unfortunately, such studies of colonisation are few and have not been carried out for a sufficiently long period. As a result, all the evidence presented for the importance of colonisation in determining rarity in the vegetation of quarries has been both parochial and circumstantial. Much work remains to be done before any categorical statement can be made about the exact nature of the factors controlling the colonisation of quarries.

## CONCLUSIONS FOR APPLIED BIOLOGISTS

If the aim of reclamation work in quarries is the establishment of low-cost, low-maintenance turf, one must take account of 3 factors: quarries are infertile; natural colonisation is slow; and more contentiously, the nature of the vegetation produced by natural colonisation is determined the ecology of the species in the surrounding vegetation.

### *Fertility*

Infertility at an appropriate level is a potential tool for low-cost management. The potential value of slow-growing species in many forms of land management has been emphasised for some years by ecologists in general and the NERC Unit of Comparative Plant Ecology in particular (eg Hunt 1975). The use of low levels of fertilisers and slower-growing plants has now been shown to be practically viable and appropriate both ecologically and financially by, amongst others, Bradshaw and his co-workers.

### *Choice of species*

Unlike most agricultural plants, most native species are adapted to infertile conditions and are therefore more suitable for much low-cost reclamation work. Some seed is now available commercially and techniques for the introduction of native species are described by Wells *et al* (1981).

### *Limestone quarries*

A calcareous quarry, unlike its acidic counterpart, is an infertile rather than a toxic habitat and a large number of species are natural colonists. Since such colonisation requires both mobility and adaptation to the quarry environment, it does not follow that the commonest quarry species listed in Table 2 are the best ones to introduce. Many other species of calcareous grassland may be more successful.

### *Siting of wilderness areas*

Areas set aside as refuges for plants and animals should be sited as close as possible to vegetation from which natural invasion of plants and animals could take place.

## ACKNOWLEDGEMENTS

I would like to thank Dr J P Grime and Dr R Hunt for permission to use jointly collected data and Professor A J Willis for his constructive criticism of the manuscript. The assistance of Miss A Pearce is also gratefully acknowledged.

## REFERENCES

BRADSHAW, A.D. & CHADWICK, M.J. 1980. *The restoration of land.* Oxford: Blackwell.

DAVID, R.W. 1981. The distribution of *Carex ericetorum* Poll. in Britain. *Watsonia,* **13**, 225-226.

DAVIS, B.N.K. 1976. Wildlife, urbanisation and industry. *Biol. Conserv.,* **10**, 249-291.

DUFFEY, E. 1973. Wildlife management in nature reserves in Britain. *Ochr. Przyr.,* **38**, 9-26.

ETHERINGTON, J.R. 1978. Eutrophication of limestone heath soil by limestone quarrying dust and its implications for conservation. *Biol. Conserv.,* **13**, 309-319.

**GEMMELL, R.P.** 1980. Derelict land as a wildlife resource. *Br. Ass. Advmt Sci. A. Mtg,* Section K. no. 189, 1-7.

**GOODMAN, G.T.** 1974. Ecology and the problems of rehabilitating wastes from mineral extraction. *Proc. R. Soc. A,* 339, 373-389.

**GRIME, J.P. & LLOYD, P.S.** 1973. *An ecological atlas of grassland plants.* London: Edward Arnold.

**HOLLIDAY, R.J. & JOHNSON, M.S.** 1979. The contribution of derelict mineral and industrial sites to the conservation of rare plants in the United Kingdom. *Miner. Environ.,* 1, 1-7.

**HOPE-SIMPSON, J.F.** 1940. Studies of the vegetation of the English chalk. VI. Late stages in succession leading to chalk grassland. *J. Ecol.,* 28, 386-402.

**HUNT, R.** 1975. Vegetation composition and plant growth-rates. *Landsc. Res. News,* 1 (10), 6-8.

**JOHNSON, M.S. & BRADSHAW, A.D.** 1979. Ecological principles for the restoration of disturbed and degraded land. *Appl. Biol.,* 4, 141-200.

**MANNING, W.J.** 1971. Effects of lime dust on the leaf condition, foliar disease incidence and leaf surface microflora of mature plants. *Environ. Pollut.,* 2, 69-76.

**NATURE CONSERVANCY COUNCIL.** 1981. Sixth Report. 1 April 1979-31 March 1980. pp.10-11.

**PIGOTT, C.D. & WALTERS, S.M.** 1954. On the interpretation of the discontinuous distributions shown by certain British species of open habitats. *J. Ecol.,* 42, 95-116.

**RATCLIFFE, D.A.** 1974. Ecological effects of mineral exploitation in the United Kingdom and their significance to nature conservation. *Proc. R. Soc. A,* 339, 355-372.

**USHER, M.B.** 1973. *Biological management and conservation.* London: Chapman and Hall.

**WELLS, T., BELL, S. & FROST, A.** 1981. *Creating attractive grasslands using native plant species.* Nature Conservancy Council.

# REGIONAL VARIATION IN QUARRIES

**B N K DAVIS**
*Institute of Terrestrial Ecology, Monks Wood Experimental Station, Huntingdon*

Quarries are extremely diverse. We should therefore look briefly at the range of conditions found in them. Detailed research studies and practical reclamation schemes are tailored to particular sites and it is important to see how widely any particular interpretation or solution may apply. The following account draws on a survey of some 200 chalk and limestone quarries in England during 1974-6. Together, chalk and limestone are among the most widely used and heavily exploited minerals in Britain (Healing & Harrison 1975; Blunden 1975). They present major problems for restoration (Barratt *et al* 1970), but many old quarries have become naturally revegetated and now escape formal classification as derelict land. Chalk and limestone quarries therefore present both a considerable challenge and some encouragement to ecologists.

Many plant species grow on calcareous substrates. A sample of 48 quarries, ranging from 15 to about 100 years old and distributed through 21 English counties, contained 428 species of vascular plants. However, the main impression was of a strong underlying similarity in vegetation, based on the prevalence of species common to all age classes, all regions and all types of limestone (Table 6). Other widespread species were restricted to early, pioneer stages of colonisation or to older, well vegetated conditions. Less common plants were mainly local or regional species but some were quite rare (Davis 1979). The latter included species dependent upon open ground conditions such as matgrass fescue *Nardurus maritimus* and plants of closed grassland communities like perennial flax *Linum anglicum*. Ancient workings such as Barnack Hills and Holes in north-west Cambridgeshire are now amongst the best remaining examples of species-rich grassland in the county. However, early successional stages can also be of botanical interest for the large population of some species that they support; eg the Chiltern gentian *Gentianella germanica* in certain Bedfordshire and Hertfordshire chalk pits.

## INFLUENCE OF QUARRYING TECHNIQUES ON TOPOGRAPHY AND THE DEVELOPMENT OF VEGETATION

### *Quarry floors*

Until the introduction of steam power, quarries were worked by hand and mainly for local needs. Progress was therefore slow and intermittent and natural revegetation of worked-out areas was able to keep pace with new workings. Such methods lingered on until quite recently and one can still see examples in which all stages from bare ground to open grassland, scrub and woodland are compressed into a short distance on the quarry floor, eg at Claxby chalk pit, Lincolnshire which is now a nature reserve. Today, hand working is reserved for a few of the finest building stone quarries used for repair work such as those at Ancaster and Holywell in Lincolnshire (Plate 1).

Many quarries have had several growth phases associated with different markets and extraction techniques. Hopton Wood quarry, near Wirksworth in Derbyshire, illustrates these phases well as the Carboniferous limestone in the area occurred in three forms (Figure 1). A fine freestone was worked for many important buildings throughout the 19th century but the quarried area was small. It was absorbed in the early 1900s within the much larger scale extraction of chemically high grade limestone (99% $Ca\ Co_3$) for lime burning. This phase left very shattered faces, much talus and large quantities of

TABLE 6 The most common plant species in 48 chalk and limestone quarries in England. The quarry sample contains 12 in each of 4 age classes, <15, 15-35, 35-55, >55 years. Total records for each species followed by number of records in youngest age class.

HERBACEOUS PLANTS

| | | | | | |
|---|---|---|---|---|---|
| *Achillea millefolium* | 35 | 5 | *Leontodon hispidus* | 34 | 7 |
| *Bellis perennis* | 28 | 8 | *Linum catharticum* | 44 | 9 |
| *Carlina vulgaris* | 22 | 3 | *Lotus corniculatus* | 43 | 9 |
| *Centaurea nigra* | 37 | 7 | *Medicago lupulina* | 35 | 8 |
| *C. scabiosa* | 21 | 5 | *Origanum vulgare* | 22 | 6 |
| *Centaurium erythraea* | 24 | 8 | *Pastinaca sativa* | 18 | 6 |
| *Cerastium holosteoides* | 24 | 8 | *Pimpinella saxifraga* | 19 | 1 |
| *Chamaenerion angustifolium* | 38 | 11 | *Plantago lanceolata* | 40 | 10 |
| *Chrysanthemum leucanthemum* | 27 | 6 | *P. media* | 18 | 1 |
| *Cirsium arvense* | 22 | 7 | *Polygala vulgaris* | 17 | 2 |
| *Clinopodium vulgare* | 18 | 4 | *Poterium sanguisorba* | 23 | 3 |
| *Dactylorhiza fuchsii* | 23 | 3 | *Primula veris* | 17 | 2 |
| *Daucus carota* | 16 | 3 | *Prunella vulgaris* | 40 | 10 |
| *Epilobium montanum* | 19 | 7 | *Ranunculus repens* | 28 | 9 |
| *Euphrasia officinalis* | 27 | 4 | *Senecio jacobaea* | 38 | 9 |
| *Fragaria vesca* | 31 | 9 | *Silene cucubalus* | 18 | 4 |
| *Geranium robertianum* | 21 | 5 | *Taraxacum officinale* | 28 | 6 |
| *Heracleum sphondylium* | 21 | 2 | *Thymus drucei* | 19 | 2 |
| *Hieracium pilosella* | 37 | 4 | *Trifolium pratense* | 25 | 3 |
| *Hypericum perforatum* | 29 | 9 | *T. repens* | 17 | 3 |
| *Inula conyza* | 28 | 10 | *Tussilago farfara* | 39 | 11 |
| *Knautia arvensis* | 18 | 1 | *Veronica chamaedrys* | 20 | 3 |
| *Lathyrus pratensis* | 16 | 1 | | | |

| GRASSES AND SEDGES | | | WOODY SPECIES | | |
|---|---|---|---|---|---|
| *Agrostis stolonifera* | 20 | 6 | *Acer pseudoplatanus* | 22 | 7 |
| *Arrhenatherum elatius* | 39 | 10 | *Betula pendula* | 16 | 5 |
| *Brachypodium sylvaticum* | 29 | 7 | *Clematis vitalba* | 22 | 6 |
| *Briza media* | 21 | 0 | *Crataegus monogyna* | 38 | 8 |
| *Carex flacca* | 23 | 2 | *Fraxinus excelsior* | 27 | 6 |
| *Catapodium rigidum* | 24 | 9 | *Rosa canina* agg. | 20 | 5 |
| *Dactylis glomerata* | 38 | 10 | *Rubus fruticosus* | 33 | 9 |
| *Festuca ovina* | 24 | 4 | *Salix caprea* | 28 | 10 |
| *F. rubra* | 36 | 7 | *Sambucus nigra* | 22 | 5 |
| *Holcus lanatus* | 38 | 9 | | | |
| *Poa pratensis* | 22 | 5 | | | |
| *Trisetum flavescens* | 31 | 7 | | | |

waste stone fragments which were tipped down the hillside like scree or heaped into large spoil banks (Plate 14). Similar results can be seen in Carboniferous limestone quarries worked in this way up till the 1950s in Derbyshire, Yorkshire, Cumbria and Somerset. The resulting topographical and microclimatic variability provides suitable conditions for many species of plants. However, the hard stone resists weathering and so a closed vegetation is slow to develop on such coarse substrates.

The third phase of development at Hopton Wood followed the enormous expansion of the aggregate market after the war, using the chemically less pure mountain limestone. More modern blasting and extraction techniques and visual sensitivities resulted in a concealed quarry driven diagonally

*Plate 1 Overburden and face of Glebe quarry, Ancaster, Lincolnshire in 1975*

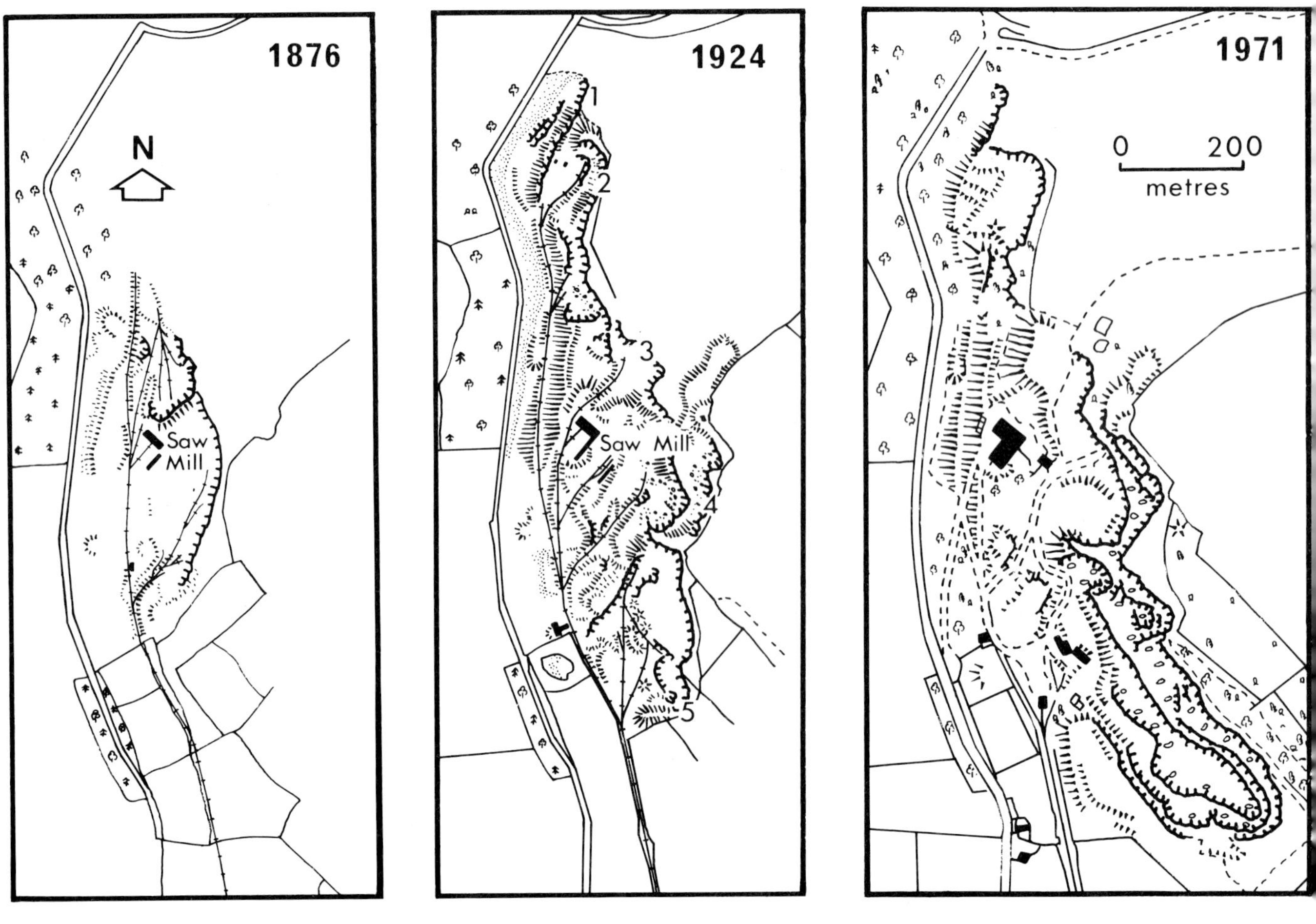


*Figure 1 Hopton Wood quarry, Wirksworth, Derbyshire 1876-1971. Crown copyright reserved.*

into the hillside in place of the earlier "bluff" quarrying along the side of the hill. The floor and terraces left in 1964 are relatively flat and free of coarse stone waste and have gained a 50% vegetation cover with an average of 20 species/$m^2$. This density of species is comparable with the much older but rougher quarry floor at the northern end of the complex which is now a nature reserve.

Similar species densities occur in some 30-40 year old chalk and Magnesian limestone quarry floors in Kent and Co. Durham (Table 7) (Plate 4). The highest density I have found was on the floor of a Jurassic limestone quarry at Clipsham, Leicestershire where work finished in 1941. In 1980, an experimental area of 150$m^2$ contained a total of 77 species at a mean density of 30.3/$m^2$ and mean cover of 21.4%. The low nutrient levels (Table 8), low annual rainfall (ca 570mm) and rabbit grazing are probably responsible for the slow development of vegetation cover here.

**TABLE 7 Total and mean numbers of plant species in 8 random $m^2$ quadrats on quarry floors with approximate dates of closure and present day adjacent land use. The quarries were virtually devoid of spoil material except at Hopton Wood (old quarry) and Ferriby Cliffe.**

| | DERBYSHIRE | | KENT | |
|---|---|---|---|---|
| | Carboniferous Limestone | | Chalk | |
| | Hopton Wood | Hartington Station | Gt. Culand | Cliffe |
| Last worked | 1920s | 1955 | 1941 | 1970 |
| Total species | 38 | 29 | 42 | 27 |
| Mean species/$m^2$ | 17.4 | 11.0 | 19.6 | 10.1 |
| Mean % cover | 66 | 27 | 76 | 91 |
| Adjacent land | grassland/wood | grassland | arable/ grassland | arable |

| | DURHAM | | | HUMBERSIDE | |
|---|---|---|---|---|---|
| | Permian Limestone | | | Chalk | Jurassic Lst. |
| | Wingate | Mill | Trimdon Grange | Ferriby Cliffe | Maws |
| Last worked | ?1947 | 1948 | 1944 | 1964 | 1934 |
| Total species | 52 | 55 | 47 | 28 | 28 |
| Mean species/$m^2$ | 19.3 | 20.9 | 15.5 | 9.8 | 12.0 |
| Mean % cover | 70 | n.d. | n.d. | 39 | 56 |
| Adjacent land | arable | quarry/ grassland | arable | arable/wood | arable |

n.d. = no data

**TABLE 8 Levels of plant nutrients in the top 5-6 cm of quarry floor at Clipsham, Leicestershire after 40 years, compared with the levels in an old pasture nearby. (Means ± standard deviations).**

| | Total % | Extractable ppm | | |
|---|---|---|---|---|
| | N | P | K | Mg |
| Clipsham quarry | 0.08<br>±0.02 | 3.2<br>±1.1 | 55.4<br>±9.6 | 79.1<br>±7.0 |
| Old pasture | 0.79<br>±0.06 | 23.5<br>±2.6 | 268<br>±124 | 213<br>±26 |

*Overburden*

The depth and nature of overburden can also influence greatly the methods of working and the final land form. In modern large quarries, such material is usually sold, utilised in some subsidiary process or used for planned landscaping of worked-out areas. Formerly, however, it was simply a waste material which had to be carried away and dumped. Often this covered up older areas of working so that little or none of the underlying rock remained exposed. Considerable depths of clay overburden were encountered in some of the Jurassic limestone quarries of Northamptonshire, Leicestershire and Lincolnshire (Plate 1). A large spoil bank at Clipsham quarry is now about 50 years old and fully vegetated with more than 100 species of plants, including 14 species of shrubs and trees (Davis 1981). Here the species density was about $20/m^2$ though this was reduced in areas of denser scrub.

In Warwickshire, the depth of sandy clay and shale overburden above the Lias limestone reached ten metres in the Harbury and Stockton area (A.P.C.M. 1952). Draglines were used in the early 1950s at Ufton to remove the clay and they created large areas of "hill and dale" like those in the ironstone workings of Northamptonshire. At Harbury, the use of a conveyor belt fed by a walking dragline produced massive spoil mounds (cf. Plate 2). The most striking vegetational features of Harbury were the exceptionally large populations of kidney vetch *Anthyllis vulneraria* and narrow-leaved birds-foot trefoil *Lotus tenuis* in the newer area and butterfly orchid *Platanthera chlorantha* in an older area. The latter two species are rare in quarries. The hawkweed *Hieracium strumosum* was abundant at Stockton. This is one of 11 species found to be early and abundant colonists of quarry spoil whether of limestone, chalk marl or clay (Davis 1977). Some of them are more common in man-made habitats such as quarries and railways than in natural habitats.

At Betchworth in Surrey, the too pure Upper Chalk had to be removed to obtain the clay-rich Lower Chalk used for building lime from the mid-nineteenth century (Searle 1935). Work continued here until 1968 and several very large spoil heaps were produced with the aid of steam power and narrow gauge railways. These have become colonised by a typical species-rich chalk quarry flora with over 80 species of plants (Plate 3). Some areas are becoming dominated by hawthorn *Crataegus monogyna* and dogwood *Thelycrania sanguinea* scrub whilst others are in various stages of development towards birch *Betula pendula* or ash *Fraxinus excelsior* woodland. As in the adjacent downland itself, this development has accelerated since the loss of rabits in the 1950s.

*Plate 2 Limestone face, spoil mound and lake at Stockton quarry, Warwickshire in 1975*

*Plate 3 Face (right) and spoil mound (left) at Betchworth quarry, Surrey in 1975*

*Water table*

Water shortage is often a limiting factor for plant growth in chalk and limestone quarries except where quarrying extends to near or below the water table. This is becoming a more common practice in large scale modern workings and will probably develop considerably in the future. There are deep pools at the Harbury and Stockton cement workings mentioned above with fringing lesser reedmace *Typha angustifolia* and other aquatic plants at the foot of the spoil mounds (Plate 2). At Ufton Fields, the hollows in the hill and dale are largely flooded and have developed a rich flora and fauna. It is now a statutory Local Nature Reserve.

At the Cliffe cement works on the Thames estuary in Kent, the chalk was excavated by face shovels to a depth of about 15m - about sea level. The pits were kept dry by pumping water to the works but the damp conditions were evident in the older disused pit from the presence of reed *Phragmites communis* and willow carr. The northern pit was abandoned in 1970 and plant colonisation here has been rapid. By 1978, most of the floor had 100% vegetation cover dominated by creeping bent *Agrostis stolonifera* and false oat-grass *Arrhenatherum elatius* with much grey willow *Salix cinerea* and a showy profusion of spotted and marsh orchids *Dactylorhiza fuchsii* and *D. praetermissa*. The species density, however, was only 10/m$^2$ (Table 2). Was this due to the lack of natural chalk species in the surrounding agricultural land or the effect of competitive exclusion resulting from the unrestricted growth of grasses?

A much more varied vegetation occurs in the similar but older chalk quarries in Essex such as Grays and Warren pits. Interesting communities are also developing in some of the Bedfordshire chalk quarries worked close to the water table. They include the unexpected occurrence of flattened meadow-grass *Poa compressa* as a dominant coloniser in damp areas (Dony, *pers. comm.*). It occurs in a similar situation in a limestone quarry near Clitheroe in Lancashire. The species is normally associated with dry banks and walls.

## SURROUNDING LAND AND SUCCESSION

Many of Hodgson's comments on colonisation and succession in quarries in the Sheffield area apply throughout the country. Salt Lake quarry in North Yorkshire has acquired, over 80 years, many of the species characteristic of the limestone in the Ingleborough area, including bird's-eye primrose *Primula farinosa* in exceptional density. Likewise, several of the older Magnesian limestone quarries of Co. Durham contain communities derived from the grassland that was formerly widespread in the area (Richardson *et al* 1980). In contrast, many active or recently worked quarries are now surrounded by agricultural land (Table 2). Both initial colonisation and succession may therefore be very different from what occurred in the past. Ferriby Cliffe chalk quarry, near Barton-upon-Humber, is surrounded by arable land and lacks 25 of the most common quarry species in Table 1. These include early colonists such as common centaury *Centaurium erythraea*, perforate St. John's-wort *Hypericum perforatum* and rough hawkbit *Leontodon hispidus*. The quarry was worked for the Scunthorpe steel industry until 1968 and large quantities of coarse chalk waste have been levelled off and left. There is now a 20-50% vegetation cover but an average of only 10 species/m$^2$ (Table 2). Wild strawberry *Fragaria vesca* is dominant. Older areas suggest that succession leads directly to hawthorn/rose scrub with ash and sycamore *Acer pseudoplatanus* without an intermediate closed grass/herb stage. Experiments would show whether dispersal or chemical/physical properties of the chalk spoil are limiting factors for the "missing" herbaceous species.

## INTRODUCED SPECIES

Several introduced plants have colonised quarries. One of the most striking examples is *Buddleja davidii*. This has spread from gardens since the war and now occurs in chalk and limestone quarries from Surrey to Norfolk and Dorset to Derbyshire. At Betchworth quarry described above, it has invaded an area of open ground and now forms an impenetrable monoculture. A similar garden escape that has established itself on quarry faces and spoil is red valerian *Centranthus ruber*. It occurs in profusion at Grays pit in Essex and in quarries in Humberside, Somerset, Gloucestershire and Co. Durham. Of the hawkweeds mentioned earlier, several species including *H. strumosum* are introductions into this country. The ability of species to establish themselves rapidly on nutrient poor and physically hostile media is useful in reclamation. We need to explore further and perhaps exploit the properties of naturally good colonists including non-native species if only as an initial phase in re-establishing vegetation.

## REFERENCES


**ASSOCIATED PORTLAND CEMENT MANUFACTURERS**. 1952. Around the works. 6. Harbury. *The Blue Circle* 6, 33-37.

**BLUNDEN, J.** 1975. *The mineral resources of Britain*. London: Hutchinson.

**BARRATT, M.S., COURTIER, D.B., HUGHES, D.O. & ROBERTSON, J.M.** 1970. Land use planning and mineral resources with special references to aggregates for the construction industry. *Proc. Symp. Geological Aspects of Development and Planning in Northern England, Sheffield, April 1970*, 48-66.

**DAVIS, B.N.K.** 1977. The Hieracium flora of chalk and limestone quarries in England. *Watsonia*, 11, 345-351.

**DAVIS, B.N.K.** 1979. Chalk and limestone quarries as wildlife habitats. *Miner. Environ.*, 1, 48-56.

**DAVIS, B.N.K.** 1981. Clipsham quarries: their history and ecology. *Trans. Leicester lit. phil. Soc.*, 72, 59-68.

**HEALING, R.A. & HARRISON, M.C.** 1975. *United Kingdom mineral statistics 1975* London: H.M.S.O.

**RICHARDSON, J.A., DAVIS, B.N.K. & EVANS, M.E.** 1980. Vegetation (e) Disused quarries. In: *The Magnesian limestone of Durham county*, edited by T.C. Dunn, 61-68. Durham: Durham County Conservation Trust.

**SEARLE, A.B.** 1935. *Limestone and its product*. London: Ernest Benn.

# QUARRIES AND NATURE CONSERVATION - OBJECTIVES AND MANAGEMENT

**C E RANSON AND J P DOODY**
*Nature Conservancy Council, Colchester and Huntingdon.*

## GENERAL PRINCIPLES

The first part of this paper (by C.E.R.) is concerned with the general principles of ecology as they relate to the development of nature conservation interest in disused quarries. The second section (by J.P.D.) discusses some of these concepts as they relate to the problems associated with the conservation of the plants and animals of the semi-natural grasslands of the magnesian limestone area of north east England.

*Why quarries?*

In Great Britain, the Nature Conservancy Council, the Royal Society for the Protection of Birds and several County Naturalists' Trusts and local authorities have some nature reserves on land derived from mining, quarrying or tipping of wastes. What is it about this sort of land that attracts these organisations, causes them to buy such sites, and spend money on management?

There are several answers to these questions:

1. The presence of nationally or locally rare plants or animals. Some of these were once widespread but are now scarce or even confined to quarries.
2. The 'natural' development of one or more types of habitat deemed to be worthy of conservation.
3. The recognition of the potential of a recently abandoned quarry for nature conservation.
4. The offer of a quarry, free, once it has been worked out.

In many instances the passage of time has allowed communities of plants and animals to develop to a stage considered desirable by man. However, further development of vegetation may proceed to a point where the attractive features are threatened with extinction, unless active steps are taken to check the succession.

*Quarries and time*

Time-scales on which these developments occur vary greatly: years, decades, centuries, even millenia. The longer periods will contain marked climatic fluctuations and their effects on vegetation. Examples of quarries from each of these time-spans are:

Grime's Graves, Breckland, Norfolk - Stone Age flint mines dug into the Chalk, the up-cast and depressions of which have a rich calcareous flora.

The Norfolk Broads in eastern Norfolk and north east Suffolk - shallow lakes created by peat digging in the 13th and 14th centuries, still retaining substantial areas of open water, reed swamps and fen, as well as carr, etc.

Barnack Hills and Holes, Cambridgeshire - medieval limestone quarries with an excellent calcareous grassland flora.

Stonesfield Slate Quarries, Oxfordshire - 15th-18th century quarries for fissile limestone suitable for roofing domestic buildings having unvegetated, mobile 'slate' heaps and calcareous grassland.

Grays Thurrock Chalk Pit, Essex - 18th-20th century chalk-diggings for lime, whiting and cement, now with a rich vegetation (see Finegan & Harvey, this symposium).

Sewardstone, Essex - calcareous wastes of the 20th century supporting a hybrid swarm of *Dactylorhiza* species.

It is questionable whether such conditions will develop in the future. Many factors operate against the leaving of man-made sites to develop plant and animal life naturally, whether they are cliffs, screes, bare rock, loose sand, slurries or irregularly-flooded quarries. Human safety, visual amenity (in its many forms) and commercial after-uses demand modification, even uniformity. Luckily we do have the legacy of the past; and from time to time operators and planners will accept nature conservation over at least part of a recently disused quarry by selling the land or relaxing restoration conditions so that a nature reserve can be established.

As well as the more conventional after-uses, such as agriculture, forestry, industry and commercial recreation, some nature reserves have been promoted by planning authorities, countryside organisations and the mineral industry itself through the Sand and Gravel Association and similar bodies.

*Quarries as nature reserves*

What distinguishes quarries from most other types of land is the combination of extreme physical features: steep and flat surfaces, coarse and fine rock residues, stable and mobile rock and spoil, very low levels of plant nutrients; also extremes of wetness and dryness, of alkaline and acidic soils, of seasonal and diurnal heat and cold, of sun and shade. These features determine what grows and lives in a particular quarry; so do the sources from which plant propagules and animal life can colonise it.

Choosing a quarry as a nature reserve means adopting one which already has high nature conservation value, or a relatively bare site with potential for natural and artificial colonisation. Diversity of soil and microclimate will vary from site to site but, as a general rule, the more variety the better so long as there are large enough areas to support viable populations of plants and animals.

The continued existence of nature conservation features in a quarry will almost always mean management; and management involves several external factors.

1. Tenure: freehold is best, a long (25+ years) unconditional lease next best, and a licence least satisfactory.
2. Access: good access over an unimpeded right of way for road and tracked vehicles is ideal, anything less has disadvantages.
3. Neighbours: just as you cannot buy a view, so you cannot buy immunity from unsympathetic neighbours. The activities of adjoining land owners and occupiers may affect the quantity and quality of water, or produce dust and other discharges which enter the reserve, while noise and other disturbances can reduce wildlife populations. On the other hand, some adjacent land uses may benefit nature conservation.
4. The public: quarries that become nature reserves have often been abandoned for several years, and legal rights may have become established through usage; discreet observation and enquiries should precede approaches to highway authorities.

Finally, quarries are a mixture of small areas of extreme sensitivity and large robust areas where nature can be allowed to take its course or vegetational succession can be checked at a particular point. If you do not like what nature has done for you, you can turn the clock back by putting in the bulldozer and letting colonisation start again, or you can import soils, seeds, plants, turves and with skill and luck produce a completely man-made nature reserve.

## APPLICATIONS IN NORTH EAST ENGLAND

Grasslands in which blue moor-grass *Sesleria albicans* is dominant occur in a zone across Northern England to the Burren in Western Ireland and form a link between the calcareous grasslands to the north and south. Those which survive in North East England are at the most thermophilous end of the range of variation and support a number of species at the northern limits of their distribution in Britain, growing together with other plants which are close to their southern geographical limits (Table 9). It is clear from Baker and Tate's Flora of Northumberland and Durham (1868) and a more recent paper by Heslop-Harrison & Richardson (1953) that the distribution of many of the species of the Magnesian limestone has been drastically reduced.

### *The contribution of quarries*

At about the same time that agricultural intensification and industrial and housing development began to take its toll of the semi-natural grasslands, a number of relatively small quarries became abandoned. These sites provided ideal conditions for the establishment and survival of a large number of plants and animals of calcareous grasslands by recreating conditions of open ground somewhat like those that existed in late glacial times when many of the species were widespread. These 'relicts' of the late glacial flora include the northern bird's-eye primrose *Primula farinosa*, dark-red helleborine *Epipactis atrorubens* and globe flower *Trollius europaeus* amongst the rarer plants, with widespread species such as rock-rose *Helianthemum chamaecistus* which are now restricted to calcareous soils.

Thus, the disused quarries of the Magnesian limestone provided a refuge for plants of the semi-natural grasslands, including many which themselves were 'relicts' of an older flora, intolerant of competition (Table 10). The commoner plants, including several orchids, and of course some of the animals may also be well represented (Richardson *et al* 1980; Dunn 1980); in the case of Durham the presence of rock-rose is important to the survival of the Durham argus butterfly *Aricia ataxerxes*.

### *Management*

It can be seen from the above that the main contribution of disused quarries for nature conservation in this area is as a reservoir for plants and animals of the semi-natural grasslands. Increasingly, as the grasslands have disappeared, the full range of species present throughout the North East has only been maintained by the substitute habitat provided by disused quarries. The primary aim, therefore, for management of these disused quarries is to retain the full range of species, particularly those associated with the grasslands. Owing to the variation in their age and the diversity of aspect and substrate type the vegetation is often very variable. In those situations, management will be aimed not only at protecting the secondary grassland development upon the quarry floor but also other vegetation including scrub, woodland and wetlands. A very delicate balance has to be achieved between the protection of the grassland species and other

**TABLE 9 The presence of some of the northern and southern elements in the semi-natural grasslands of north east England.**

MORE IMPORTANT SPECIES WITH A GENERALLY NORTHERN DISTRIBUTION

| SPECIES | STATUS |
|---|---|
| *Sesleria albicans* | Widespread and abundant in many of the surviving grasslands but absent from much of the coast. |
| *Epipactis atrorubens* | Now only known from one semi-natural site (Thrislington Plantation) and one roadside verge. |
| *Primula farinosa* | Much reduced; present probably only at Town Kelloe Bank SSSI and Cassop Vale SSSI. |
| *Antennaria dioica* | Probably never common but now only known from one small part of Thrislington Plantation. |
| *Cirsium heterophyllum* | Present in one of the coastal denes. Although plentiful in the west of Durham virtually absent from the Magnesian limestone. |
| *Pinguicula vulgaris* | Much reduced but still present in a few sites (Town Kelloe has the best population). |
| *Trollius europaeus* | Reduced, now only definitely known from Cassop Vale and one other small site near Sherburn. |
| *Dactylorhiza purpurella* | Still present in one or two sites including Thrislington Plantation. |

MORE IMPORTANT SPECIES WITH A GENERALLY SOUTHERN DISTRIBUTION

| | |
|---|---|
| *Linum anglicum* | Present only at Thrislington Plantation and two small sites in Tyne and Wear. Extinct from a number of sites. |
| *Ophrys apifera* | The only known extant native site is on coastal grassland near Hawthorn Dene. Undoubtedly extinct in a number of former grassland localities. |
| *Ophrys insectifera* | Probably never common but now only found in shaded grassland in Castle Eden Dene. |
| *Hypericum montanum* | Now only present in small areas of grassland on the Durham Coast. |
| *Aquilega vulgaris* | Only one or two surviving possibly native sites including Sherburn Hill. |
| *Zerna erecta* | Present in a number of sites growing together with blue moor-grass. |
| *Serratula tinctoria* | Probably now restricted to the Durham Coast where it is quite abundant. |
| *Anacamptis pyramidalis* | Very few localities inland (Cassop Vale and Hasting Hill) although quite common on the coast. |
| *Blackstonia perfoliata* | Probably no native sites survive. |

**TABLE 10 The main quarries and their contribution to the conservation of some of the less common plants of the semi-natural grassland of north east England.**

N and S denote species of predominantly northern or southern distribution

* Sites which are partly or entirely reserves of the Durham County Conservation Trust.

| | | | |
|---|---|---|---|
| * Bishop Middleham SSSI | *Epipactis atrorubens* | N | Despite major reworking of the southern of two quarries, one of the richest sites in the North East with a wide variety of grassland plants and animals. |
| | *Sesleria albicans* | N | |
| | *Anacamptis pyramidalis* | S | |
| | *Hypericum montanum* | S | |
| Wingate Quarry LNR | *Sesleria albicans* | N | |
| | *Primula farinosa* | N | |
| | *Pinguicula vulgaris* | N | |
| * Raisby Quarry SSSI | *Epipactis atrorubens* | N | A small long-disused quarry with a vigorous population of the orchid. |
| | *Sesleria albicans* | N | |
| Fulwell Quarry SSSI | *Ophrys apifera* | S | Two small colonies present of *O. apifera; B. perfoliata* plentiful. |
| | *Blackstonia perfoliata* | S | |
| Old Cassop (part of SSSI) | *Pinguicula vulgaris* | N | |
| Garmondsway Moor | *Anacamptis pyramidalis* | S | The only site in the North East for *C. eriophorum* |
| | *Blackstonia perfoliata* | S | |
| | *Cirsium eriophorum* | S | |
| Aycliffe Quarry | *Ophrys apifera* | S | One of two colonies destroyed in the western quarry by tipping. Site in the active eastern section extant. |
| * Trimdon Quarry | *Ophrys apifera* | S | Transplanted from a threatened colony at Aycliffe. |

requirements and constraints. In this context it is interesting to look at the development of Wingate Quarry (Plate 4).

This large quarry complex lies close to Wheatley Hill and consists of several connecting quarries which have been abandoned for varying lengths of time. As a consequence of this the vegetation in each is different, ranging from young ash woodland over hawthorn scrub and open invading scrub, to tall herb communities and open vegetation on the quarry floor.

Clearly, each vegetation type has its characteristic range of species, but the most important are the tall herb communities which are particularly rich in insects, and the open communities of the quarry floor which include three species uncommon in semi-natural grassland, blue moor-grass, bird's-eye primrose and butterwort *Pinguicula vulgaris*, all representative of the northern element of the flora.

In 1975, when the Nature Conservancy Council first began discussion about this site with the County Council, one of the quarries had already been filled with rubbish. Whilst discussions with the County Council on tipping into another of the quarries were taking place, proposals for a shooting range and associated "war games" were dropped. Complications - over provision of money, definition of what constituted "reclamation" eligible for

*Plate 4 Part of Wingate Quarry, Durham in 1975. A rich botanical site and haunt of the Durham argus butterfly until filled with refuse.*

grant from the Department of the Environment, vandalism, access, motor-cycle scrambling and, not least, the need to convince councillors and local inhabitants of the value of nature conservation - forced a number of compromise management decisions to be made.

The first of these compromises centred around the decision by the Local Authority to use what was at that time botanically the richest part of the site for tipping domestic refuse. Attempts were made to transplant some of the more important areas of vegetation using voluntary labour. This was only partially successful since only a very small area was involved, and some 5 years later it appears that moonwort *Botrychium lunaria* may not have survived.

Further loss of vegetation ensued when part of the quarry was "reclaimed" in order to enable the Department of the Environment to give a grant to the Local Authority. The Health and Safety at Work Act then caused the closure of part of a proposed nature trail and the loss of vegetation when an alternative route was formed. Both these developments were justified because they provided the opportunity to recreate the conditions which must have existed when the quarries were first abandoned. A further area was relinquished to motor-cycle scramblers although this was already largely devoid of vegetation. Despite these losses, the declaration of the site as a Local Nature Reserve by the Local Authority in 1980, under Section 21 of the National Parks and Access to the Countryside Act, provided an opportunity to protect much of the conservation interest of the site.

Future management will include controlled use for teaching, and interpretative facilities for casual users. This, together with vegetation control by scrub removal (often a problem in ungrazed Magnesian limestone grasslands), should ensure the continued survival of not only the remaining northern elements of the flora of the semi-natural grasslands, but also a wide variety of their typical plants and animals.

It is important to realise that the diversity of the flora of these quarries in no way matches that of the more important, much older semi-natural grasslands. Discussions over quarrying proposals at Thrislington and agricultural activities at Cassop Vale may lead in the long-term to the protection of the more important parts of these, the only nationally significant examples of semi-natural grasslands which have developed on the Magnesian limestone. However, if the full range of species is to survive, the quarries have an important role to play, and this is perhaps reflected in the extent to which several of these have become reserves of the Durham County Conservation Trust.

## REFERENCES

**BAKER**, J.G. & **TATE**, G.R. 1868. A new flora of Northumberland and Durham *Natural History Trans. of Northumb.*, 11, 1-316.

**DUNN**, T.C. & **SHEPPARD**, D.A. 1980. Insects. In: *The Magnesian limestone of Durham county*, edited by T.C. Dunn, 69-71. Durham: Durham County Conservation Trust.

**HESLOP-HARRISON**, J.W. & **RICHARDSON**, J.A. 1953. The Magnesian limestone area of Durham and its vegetation. *Trans. nth. Nat. Un.*, 2, 1-27.

**RICHARDSON**, J.A., **DAVIS**, B.N.K. & **EVANS**, M.E. 1980. Vegetation. (e) Disused quarries. In: *The Magnesian limestone of Durham county*, edited by T.C. Dunn, 61-68. Durham: Durham County Conservation Trust.

# NATURAL PROCESSES

## PLANT DISPERSAL AND COLONISATION

**HILARY GRAY**
*Department of Botany, University of Liverpool*

The many and varied industrial waste heaps that abound in Greater Manchester are not associated with quarries, but the processes of plant dispersal and colonisation on them are probably very similar.

### THE SPECIES CONCERNED

*Leblanc wastes*

The Leblanc process for making sodium carbonate, which became obsolete in 1920, produced wastes consisting largely of calcium carbonate and sulphide. This material now has a surface pH of 7.5-8. It has developed about 50% vegetation cover, mostly weed species - in particular those weeds which will colonise any piece of disturbed ground, however infertile, such as the hawkweed *Hieracium umbellatum* and creeping thistle *Cirsium arvense* (Greenwood & Gemell 1978). Such species are abundant in the adjacent communities found on roadsides, demolition sites etc, characteristic of the urban fringe. The flora of the wastes thus provides an example of the principle that whatever grows next to an area has the best chance of colonising it.

However, the heaps also support a few species characteristic of lime-rich habitats which are uncommon or absent from the region as there are no natural calcareous outcrops. Most spectacular are the large colonies of orchids (*Dactylorhiza* spp, and *Gymnadenia conopsea*). Most of these species have tiny, wind-distributed seeds. The prevailing wind is westerly, and in two cases (early marsh-orchid *Dactylorhiza incarnata coccinea* and creeping willow *Salix repens argentea*) it is the coastal subspecies which has colonised the heaps. Hence it seems likely that all the calcicole species have spread from the Lancashire sand dunes, 30-40 km away, rather than any limestone source. A similar situation exists on the calcareous Solvay wastes in Cheshire (Lee & Greenwood 1976).

*Colliery shales*

In contrast to the Leblanc wastes, the colliery tips of Greater Manchester are very acid (pH 2.5-5), owing to the pyritic shales of the Lancashire coalfield. The older tips have again developed a flora largely of invasive weeds, including some species found on the Leblanc wastes, such as rosebay willow-herb *Chamaenerion angustifolium* and the hawkweed *Hieracium umbellatum*. There also occur a few acid grassland species like wavy hair-grass *Deschampsia flexuosa* and mat-grass *Nardus stricta* and an occasional heath species such as heather *Calluna vulgaris*. This is an area of acid soils so the paucity of such colonisers is surprising. However, by the time these heaps were tipped in the 20th century, agricultural drainage had destroyed most of the local mosslands, and intensive farming was removing unimproved grasslands in the lowland areas. Seed sources to invade the tips were therefore never plentiful. In addition, most such species have relatively large seeds with poor dispersal, so even a few kilometers between the source and a heap would represent a considerable barrier.

Introduction experiments have confirmed that distance is the factor which excludes species with low powers of dispersal. Table 11 shows a selection which have established from seed and/or transplants.

**TABLE 11 Some successful introductions to industrial waste heaps in Greater Manchester**

| Alkaline wastes | success rate | Acid wastes | success rate |
|---|---|---|---|
| *Anthyllis vulneraria* | high | *Anthoxanthum odoratum* | medium |
| *Briza media* | medium | *Galium saxatile* | low |
| *Fragaria vesca* | medium | *Lotus corniculatus* (acid-tolerant clones) | high |
| *Poterium sanguisorba* | high | | |
| *Rhinanthus minor* | high | *Potentilla erecta* | low |

## SELF-FERTILITY

Baker (1955) postulated that self-fertility would be an advantage to a species that colonised over a long distance because populations could be formed from solitary individuals. Among the long-distance colonisers of this study, self-fertility is prevalent (about ¾ of the species). Most of the species that are self-sterile are limited to few heaps and small populations (eg heath bedstraw *Galium saxatile* and quaking grass *Briza media*); an extreme example is the dioecious creeping willow *Salix repens*, which has a solitary plant on each of six heaps, growing well but unable to reproduce. There is one exception, blue fleabane *Erigeron ager*, a self-incompatible annual which occurs in abundance on six heaps from St Helens to Bury. Its nearest source is Southport dunes and its success stresses that self-fertility is merely an aid; dispersability of propagules is the most important factor. The frequency of self-fertility among these species partly reflects the same in the whole sand dune flora.

## FUTURE COLONISATION

The most obvious group of species which might be expected to colonise waste heaps, but have as yet not done so, are those of fertile disturbed ground, especially fields. Since a few do occur on paths and other relatively fertile places, it is likely that they are excluded only by infertility of the substrate. All the highly mobile species from the dunes (apart from a few rare ones such as bee orchid *Ophrys apifera*) have already colonised. Indeed there are records of them being established 25 years ago on one heap, which suggests that most species which are going to cover the distance do so fairly quickly. For the acid sites, seed sources are now so few, and the species themselves so immobile, that chances of future colonisation without human intervention are tiny. Future colonisation may not be able to add much to the floristic interest of the heaps.

## ISLAND BIOGEOGRAPHY

McArthur and Wilson (1967) postulated that the number of species on an island attains an equilibrium dependent on the balance between immigration and extinction. For a given type of island in a reasonably compact geographical region, they predict a straight line relationship between the logarithms of area and species number, with a positive slope of 0.2-0.35. This model has been found applicable to many true island systems, and to 'habitat islands' on continents.

From the point of view of colonising plants, waste heaps bear certain resemblances to islands set in a solid sea. The wastes differ from their surroundings in soil chemistry and fertility, and physically in such properties as water supply and exposure. There are some problems in applying island theory to these heaps - the number of sites is small, their sizes well below anything the model has been applied to previously, and their communities early in a primary succession, when only a quasi-equilibrium of easily-dispersed species is expected.

The Leblanc sites fit the model well - correlation coefficient r = 0.87; the slope (0.16) is a little low as is usually found with habitat islands carrying many transient species. Only one site is more diverse than expected, the result of a uniquely varied tipping topography (Figure 2).

Wooded colliery heaps similarly follow the expected pattern, though with a somewhat steep slope (0.42). Interestingly, three heaps artificially planted some 25 years ago fit perfectly with the naturally colonised, older heaps, showing just how effective tree cover is at encouraging establishment.

Since unwooded colliery heaps are more common than other waste types, it seemed worthwhile to compare species number with a range of factors. No relationship between species number and colliery closure date or pH can be found. This contrasts with work on the Yorkshire coalfield which has shown a linear relationship between species number and pH. However, the Yorkshire coalfield shales range from acid to neutral, allowing many more species at the higher pH's than can tolerate the acidic Lancashire shales.

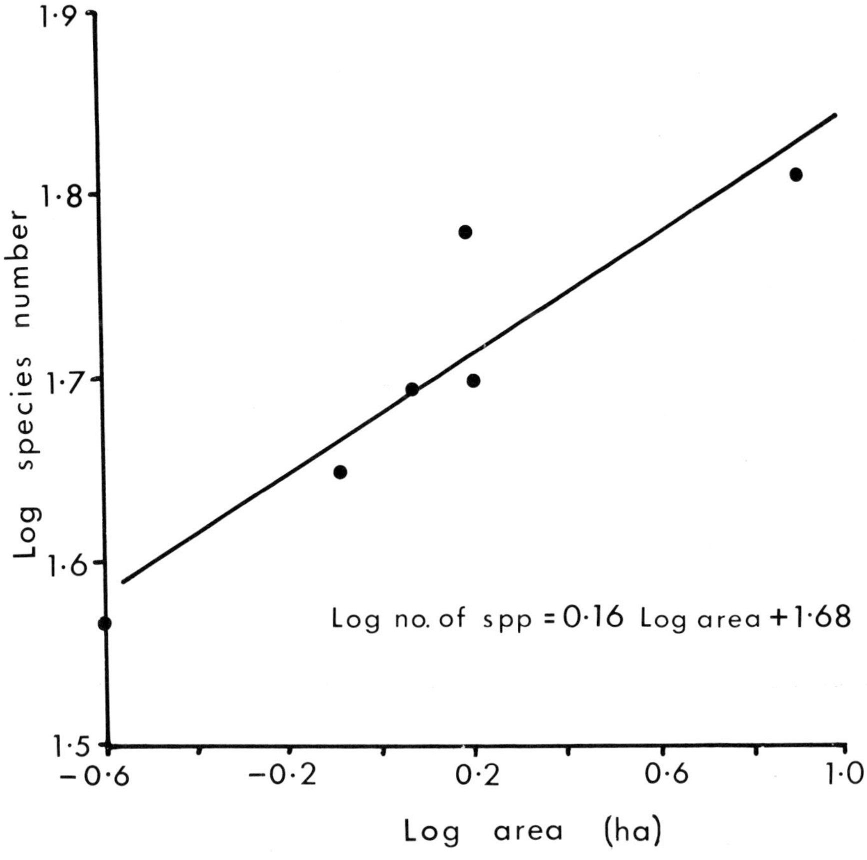


*Figure 2 Leblanc wastes: regression of log species number against log area*

When species number is compared to area (Figure 3), some effect of age appears in addition to that of area. Three sites (o) are over 100 years old, with a low, hummocky tipping pattern, and are relatively well vegetated. Two (x) are less than 30 years old and as yet little colonised. The regression line is for the main group of closures (30-50 years ago), and again shows a high slope (0.71). The great differences in age needed to produce different communities show just how slowly succession proceeds on these inhospitable shales. The high slope values are unexpected. These sites are at an early stage in succession, with competition probably limited. Immigration is greatly influenced by the number of propagules arriving and likely to be more area-dependent than is extinction. While this might explain the great effect of area on the colliery communities, it does not explain why the Leblanc wastes should show a much less marked effect.

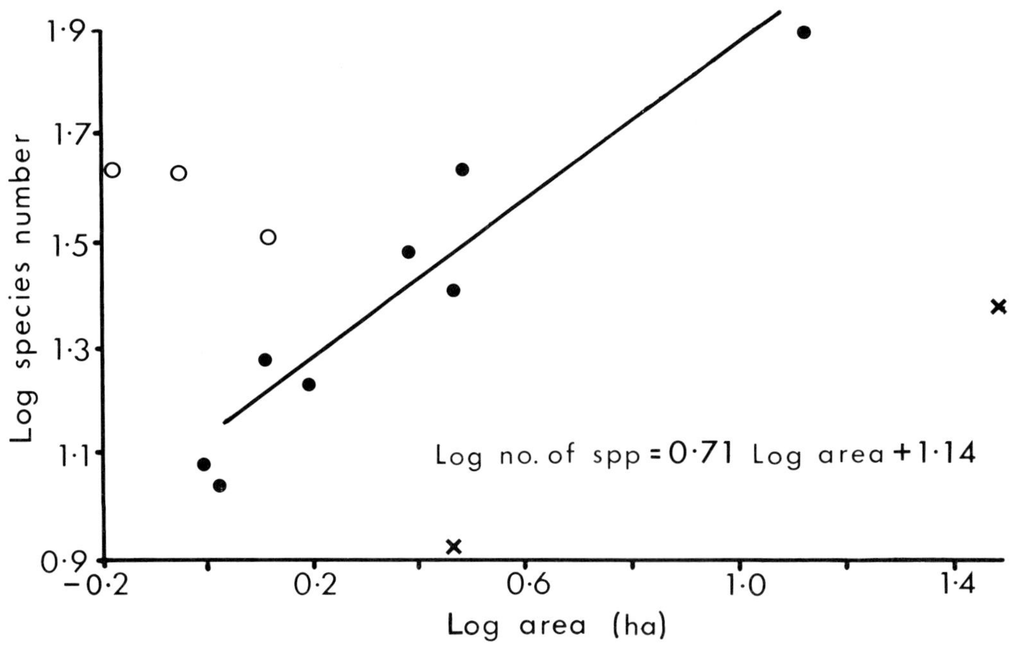


*Figure 3 Colliery shales: regression of log species number against log area*

IMPLICATIONS FOR MANAGEMENT

Because of the deficiency of floristically interesting sites and public open space in the region, it is hoped that many of the heaps examined in this study will become amenity and ecological teaching areas. For these purposes it will be necessary to considerably improve both the appearance and diversity of the sites. As shown above, succession is not likely to have any rapid effect in this direction. Tree and shrub planting may be valuable to increase the range of habitats available and the number of people an area can absorb, but this is likely to break up small areas into even smaller parcels. This in turn may decrease the number of species the areas will support - with the risk of losing the more interesting ones, and of having continually to reintroduce species to maintain an attractive community. Any planting schemes should therefore be designed not to unnecessarily fragment habitat areas.

REFERENCES

**BAKER, H.G.** 1955. Self-compatibility and establishment after 'long-distance dispersal'. *Evolution, Lancaster, Pa.*, **9**, 347-349.

**GREENWOOD, E.F. & GEMMELL, R.P.** 1978. Derelict land as a habitat for rare plants in S. Lancs (Vc 59) and W. Lancs (Vc 60). *Watsonia*, **12**, 33-40.

**LEE, J.A. & GREENWOOD, B.** 1976. The colonisation by plants of calcareous wastes from the salt and alkali industry in Cheshire, England. *Biol. Conserv.*, **10**, 131-149.

**MCARTHUR, R.H. & WILSON, E.O.** 1967. *The theory of island biogeography.* Princeton, N.J.: Princeton University Press.

# SEEDLING DEMOGRAPHY IN QUARRY HABITATS

**DAVID G PARK**
*Institute of Terrestrial Ecology, Monks Wood Experimental Station, Huntingdon*

## INTRODUCTION

Seed and seedling ecology have been largely overlooked in the study of quarries, but are areas of importance when considering the natural colonisation of these habitats. Quarrying initiates a primary succession with the removal of existing vegetation, the mineralised soil layer and the seed bank. Colonisation by plants of these denuded areas will be dependent on the influx of seed/propagules from external sources and the actual rate of colonisation will be dependent upon the outcome of seedling establishment and subsequent seedling survival.

A study has been made of both natural and experimental seedling populations of several common plant colonists. Particular attention has been given to rates of recruitment and mortality, and to ways in which these rates may be modified by experimental treatments. Results from this work will be reported in the paper.

## GENERAL ASPECTS OF SEEDLING ECOLOGY

Although the natural colonisation of disused quarries has produced a rich diverse flora, a long developmental period is often required before an appreciable cover of vegetation is established. The community remains open with much bare ground. In such a situation, the opportunity for recruitment to seedling population is likely to be greater than that in a closed community where recruitment depends on the occurrence of gaps in the vegetation cover. However, seedling mortality must be severe otherwise this open community structure would not be a persistent feature of the quarry floor habitat.

Rates of mortality among juvenile plants are very high: the young seedling is the most susceptible phase in the ontogeny of the individual and mortality generally decreases with age. This can be demonstrated in the following simple way.

1. If an individual's ontogeny is divided into a number of discrete intervals by separating either developmental stages or age states (Figure 4A), it is possible to assign probability values (Pi) to the chances of passing from one stage/state to the next.
2. Multiplication of the interval probabilities (Pi) gives the overall probability that an individual reaches a certain stage of development or age (Figure 4B).
3. Probability histograms for the Pi's for a generalised perennial and an annual show two alternatives (Figure 4C). In the first, a postulated increase in interval probability values with age/development gradually levels off. In the second, the postulated increase continues until a point is reached, perhaps coincident with flowering, when probability values decline sharply.
4. Consider the shaded areas of the histograms covering seedling establishment up to a limit of 60 days; the calculated probability of an individual passing from emerged seedling to a sixty-day old seedling is of the order 0.18-0.75, depending on species, for seedlings on quarry floors (Figure 4D). Similar work in America gives a slightly lower range of 0.08-0.55 (Skaller 1977).

A.

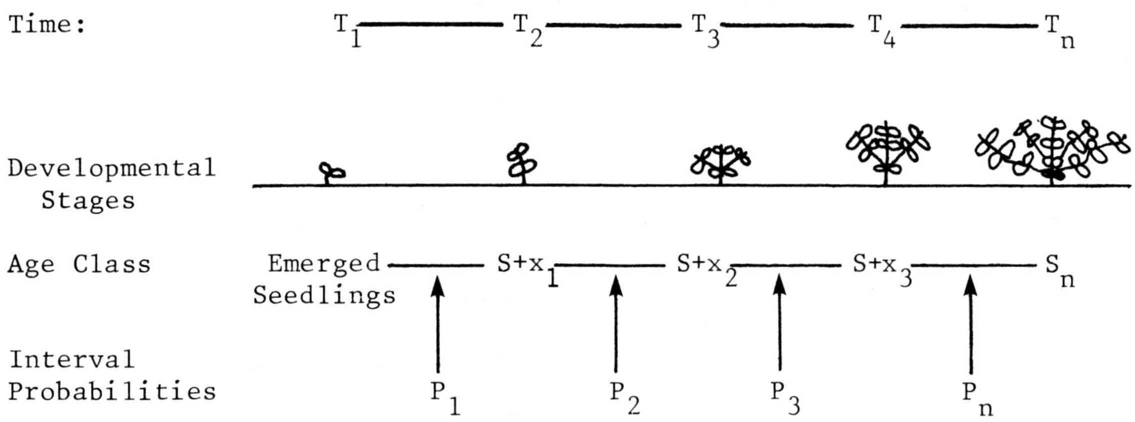


B. pr(Emerged Seedling $\longrightarrow S+x_3) = P_1 \times P_2 \times P_3$

C. Probability Histograms - perennial, annual

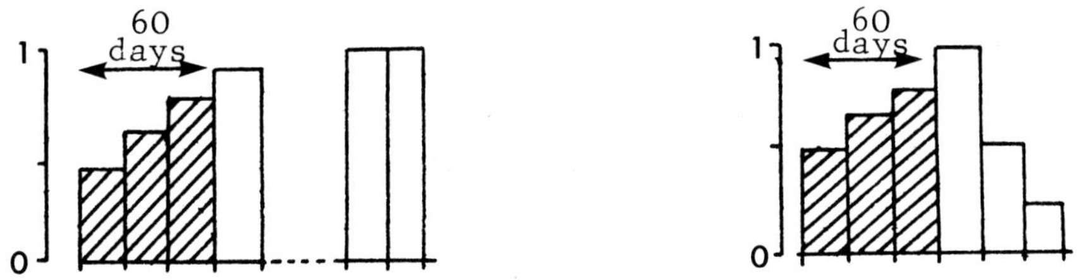


D. Quarry floor species: pr(Seedling $\longrightarrow$ S+60 days) = 0.18-0.75

E.

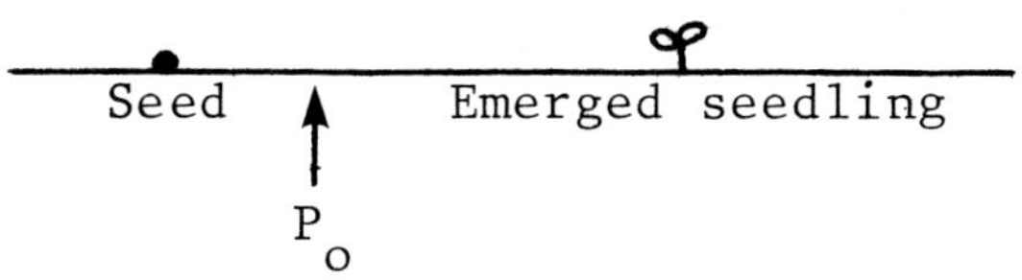


Quarry floor species: pr(seed $\longrightarrow$ seedling + 60 days) = 0.02-0.10

*Figure 4 Survival probabilities in seedlings*

5. However, if the interval from seed to emerged seedling is also included, then the probability that an individual passes from seed to sixty-day old seedling falls abruptly to 0.02-0.10 (Figure 4E).

From this it would appear that the major limiting period of natural colonisation is the passage from seed to seedling. For the particular quarry studied by Skaller, it was concluded that seed input was not a limiting factor in the colonisation process but rather the events following the arrival of seeds on the spoil surface.

## COHORT VARIABILITY

The approach outlined above, while showing the magnitude of seed/seedling loss, is an over-simplification and does not show one important aspect of seedling ecology - that of cohort variability. Baskins & Baskins (1972) found that early germinating cohorts of *Leavenworthia stylosa* suffered high mortality but had a greater probability of flowering, whereas later germinating cohorts avoided such severe early mortality but had a lower probability of flowering. This general relationship can be found in quarry seedling populations but an overall influence is exerted by the prevailing climatic conditions. Seedling emergence and mortality from a permanent quadrat established on a quarry floor, was studied intensively for two years and seedling survivorship curves were constructed for different cohorts (Figure 5). During 1979, germination was curtailed by the dry conditions occurring during June/July and the relationship between early and late cohorts is not shown. However, during 1980 this relationship can be clearly seen.

One of the most noticeable aspects of seedlings emerging in quarry habitats is the long period of time they spend at a small size and at an early stage of development (at the cotyledon, first and second leaf stages). Research elsewhere has shown that annual growth rates are small and it is common to find annuals and perennials remaining as basal rosettes for extended periods of time (Raynal 1979, Klemow & Raynal 1981). Perhaps the single most important characteristic of juvenile plants that influences their fitness is their size during environmental stress. Mortality is strongly dependent on size with smaller individuals suffering greater hazards. For many plants the total length of the juvenile period will be a function of size - itself a function of growth rates which reflect, amongst other things, nutrient availability.

## CAUSES OF SEED AND SEEDLING LOSSES

1. The most important causes of seed losses during the early stages of colonisation are probably physiographic processes such as wind and water erosion and frost heave, leading to seed burial. In an experiment on the recovery of dyed seeds sown on to a bare quarry floor, losses amounted to 20-50% after 25 days, 30-75% after 50 days and 60-90% after 125 days (which included an overwintering period). Seeds were found to have moved down the spoil profile during the course of the experiment.

2. Probably the most important factor causing seedling mortality is desiccation acting both directly, during periods of drought, and indirectly, for example after spoil movement and exposure or roots. The spoil is freely drained and the surface dries relatively quickly: during late spring and summer the moisture content of the spoil is frequently below the permanent wilting point (Figure 5).

3. Failure of the radicle to penetrate the spoil, as a result of the formation of a cementation layer or because of waterlogging of the spoil.

4. Predation or grazing by invertebrates can cause high mortality in seedling populations and may be selective in nature.

5. High surface temperatures may cause death directly and are certainly capable of stress through effects on evapotranspiration.

6. In active quarries, the adverse effects of dust can be important: dust on leaves reduces photosynthesis, interferes with transpiration and increases the risk of disease. The overriding influence of nutrient deficiency may increase the effect of these mortality agents, for example plants become more susceptible to water stress.

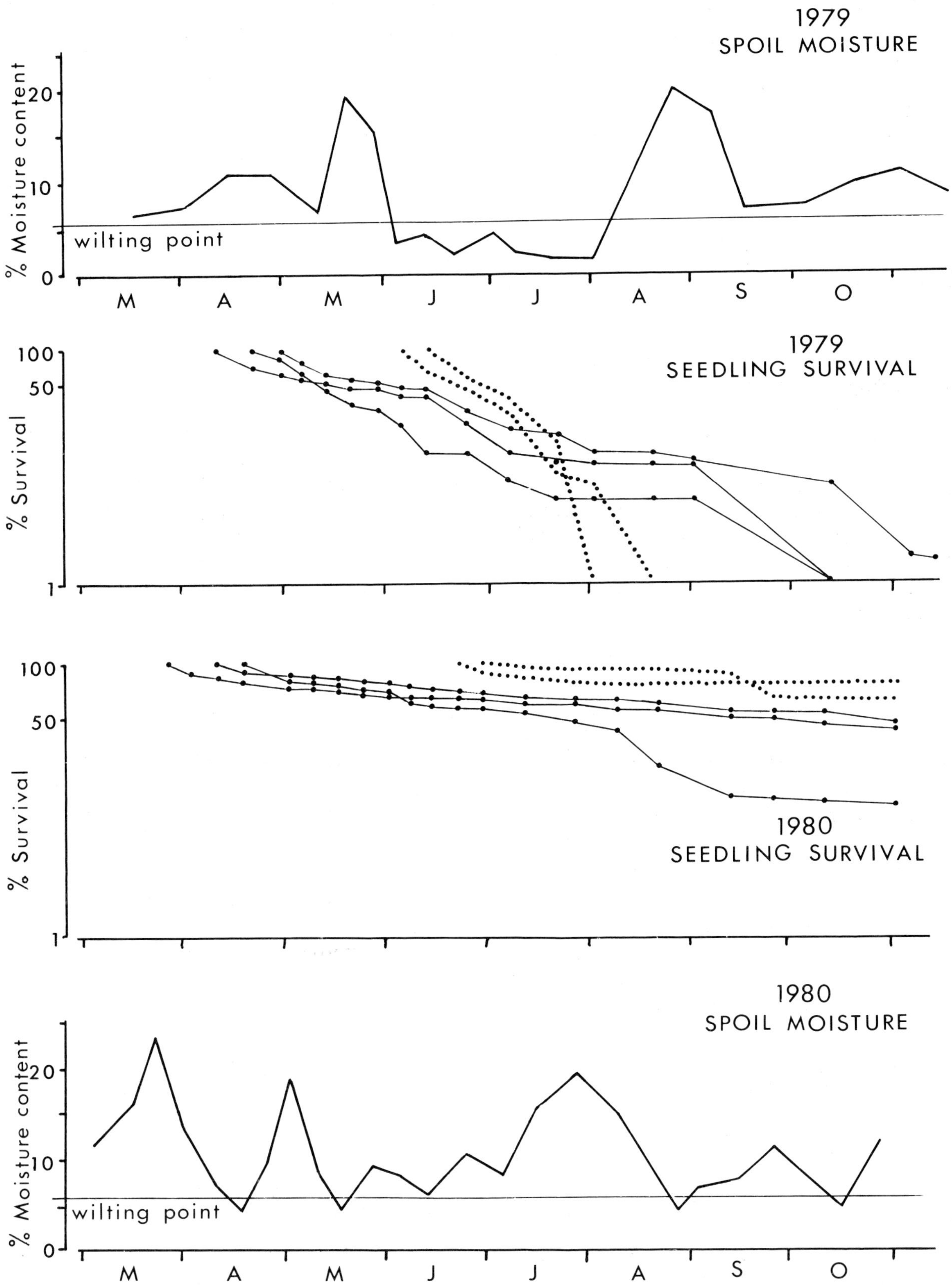


*Figure 5 Spoil moisture and seedling survivorship curves for 1979 and 1980; data from a permanent quadrat study. Survivorship curves are for* *__total__* *seedlings emerging*

## EXPERIMENTS ON *ORIGANUM VULGARE*

Some of these points may be illustrated by reference to a particular species, studied in a quarry habitat. *Origanum vulgare* (marjoram) is a common constituent of quarry floras: it is an aromatic perennial herb with a bushy appearance. The seeds are small ($6.4 \times 10^{-5}$g) and produce seedlings of only a few millimetres in size. During February 1980, a number of areas of quarry floor were artificially seeded, and seedling emergence, survival and development were monitored intensively throughout the year. The quarry floor consisted of a shallow layer (0-10cm) of crushed limestone or spoil overlying the baserock: the spoil contained little or no organic material, was freely drained and poor in both macro- and micro-nutrients.

Germination in the plots began in April and was more or less complete by July; total plot germination was in the range 18-51% and compared with a laboratory germination value of over 90%. The population flux of one of the seeded areas is illustrated (Figure 6A) and shows cumulative germination, cumulative mortality and the number of seedlings present. The seedling population peaks in early May and then declines to a level which remains relatively stable throughout July-October. Cohort survivorship curves (Figure 6B) clearly demonstrate the relationship between early and late cohorts remarked upon earlier. However, during the following winter, the population was greatly reduced by the effects of frost heave, with only 1-3% of the seedlings surviving to March 1981.

During the monitoring of the plots, the seedlings present were classified according to six stages of development (Figure 7). The histograms in this figure give the proportional representation of these classes at the dates indicated. Histogram 8 (November) shows that even after a considerable period of time (around 180 days) most of the seedlings are still at an early stage of development. Age-class frequency distributions of the seedlings are also shown for selected dates (Figure 8). The developmental stages do not reflect age classes. The size of an individual, which often reflects its stage of development, is more important in determining its behaviour than its chronological age (Gross 1981, Werner 1975).

*O. vulgare* was also used in experiments designed to increase recruitment and reduce mortality. A sawdust mulch was applied to sown plots of *O. vulgare*. Recruitment in plots mulched with sawdust was very much higher than in non-mulched plots and much of the early seedling mortality was avoided (Figure 9). The main effect of the mulch was to increase the spoil moisture 2-4%. Although recruitment and seedling survival were found to be greater in mulched plots, the rate of seedling development was unaffected and was similar to that of non-mulched plots.

## CONCLUSIONS

Seedling ecology forms an important part of the study of a primary succession such as that occurring in disused quarries. The seedling represents one of the most susceptible and vulnerable stages during the ontogeny of a plant; rates of recruitment and mortality in seedling populations may influence the rate at which colonisation proceeds.

## ACKNOWLEDGEMENTS

This work was funded by a Natural Environment Research Council Institute award. Dr B N K Davis and Mr R N Humphries gave valuable help in the planning of the research programme and in the preparation of the manuscript.

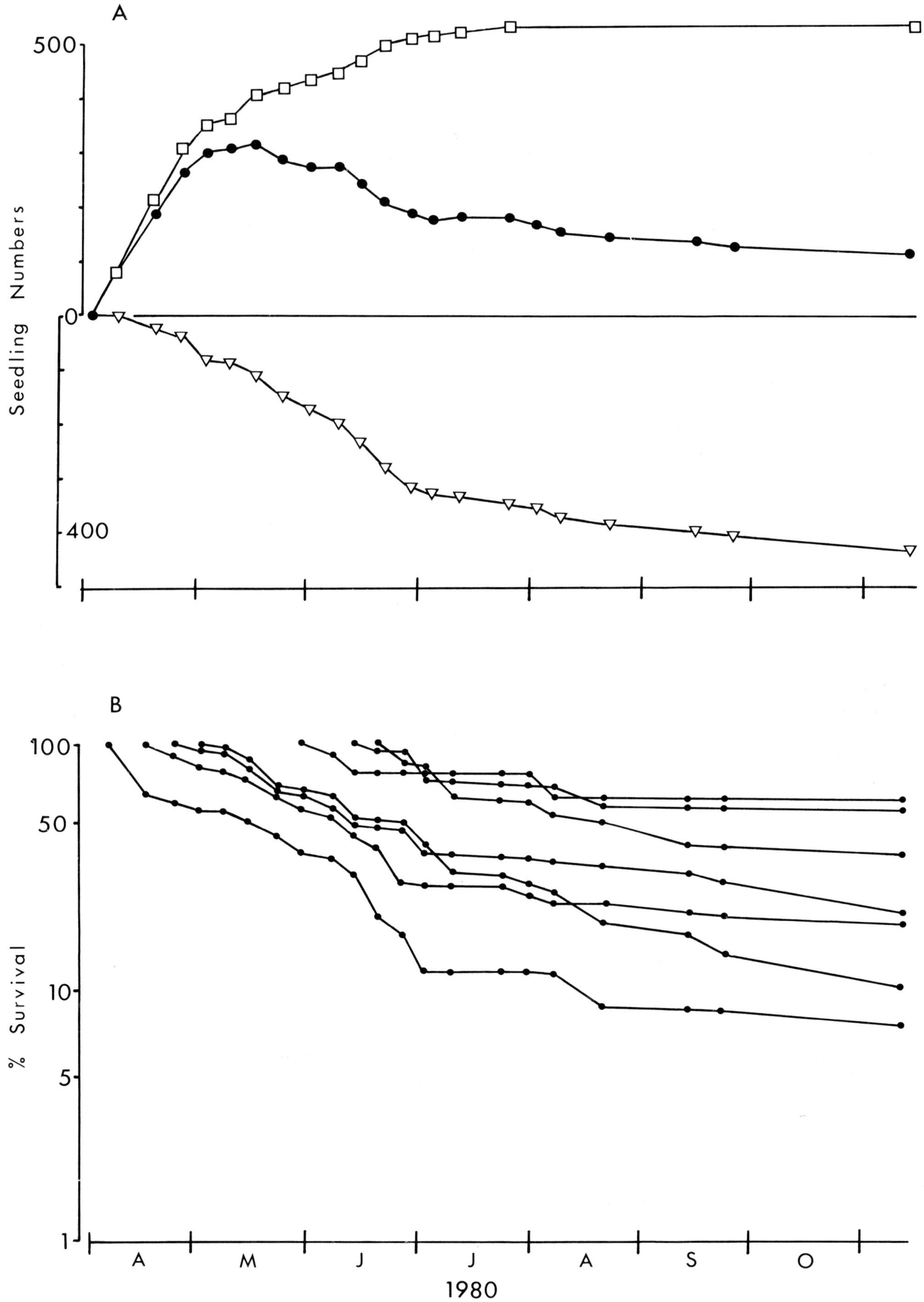


*Figure 6* **Origanum vulgare**: seedling emergence and mortality. A: population flux diagram □—□ cumulative recruitment, ▽—▽ cumulative mortality, ●—● seedlings present. B: cohort survivorship curve.

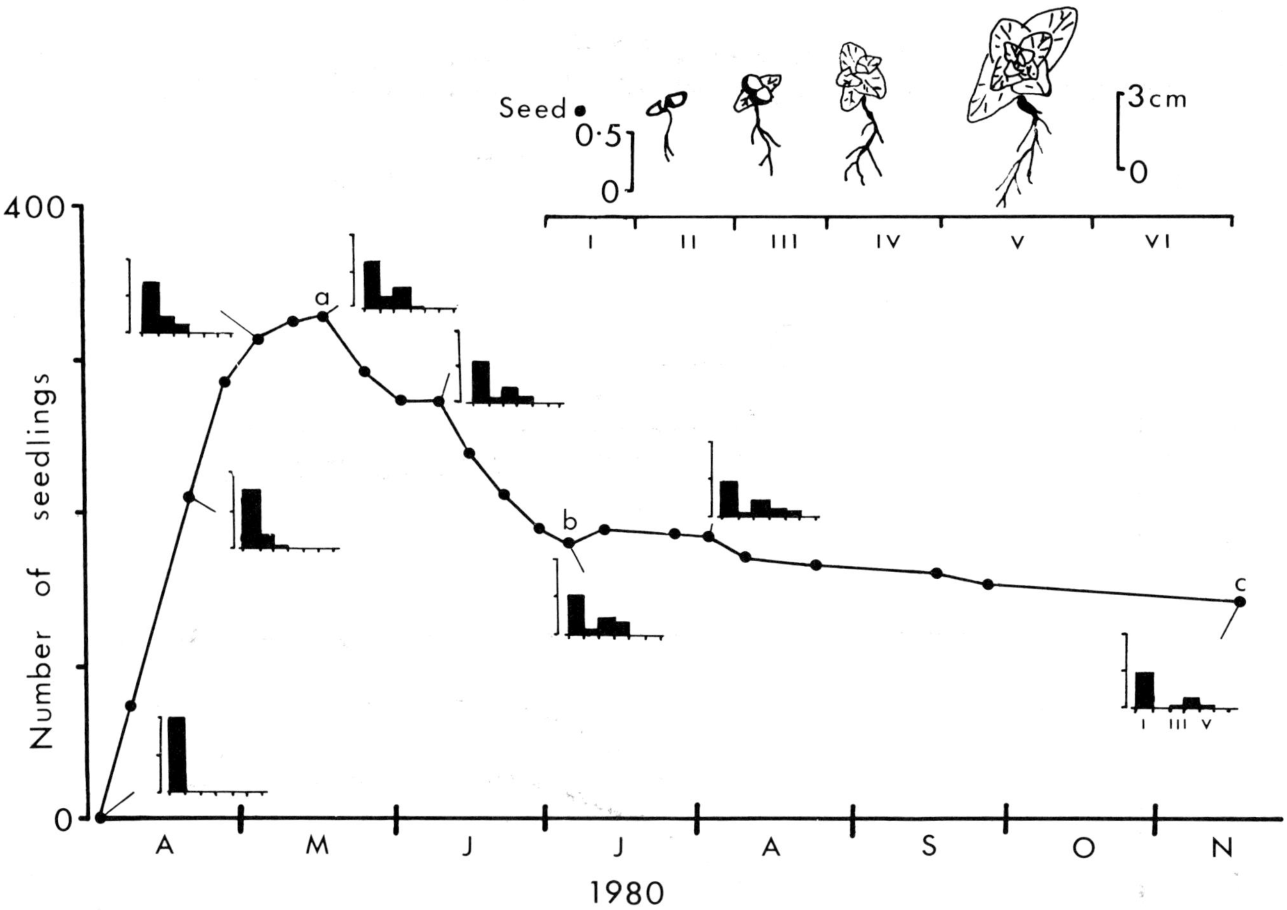


*Figure 7* **Origanum vulgare**: *Developmental stage analysis (as a proportion of seeds sown) at selected dates.*

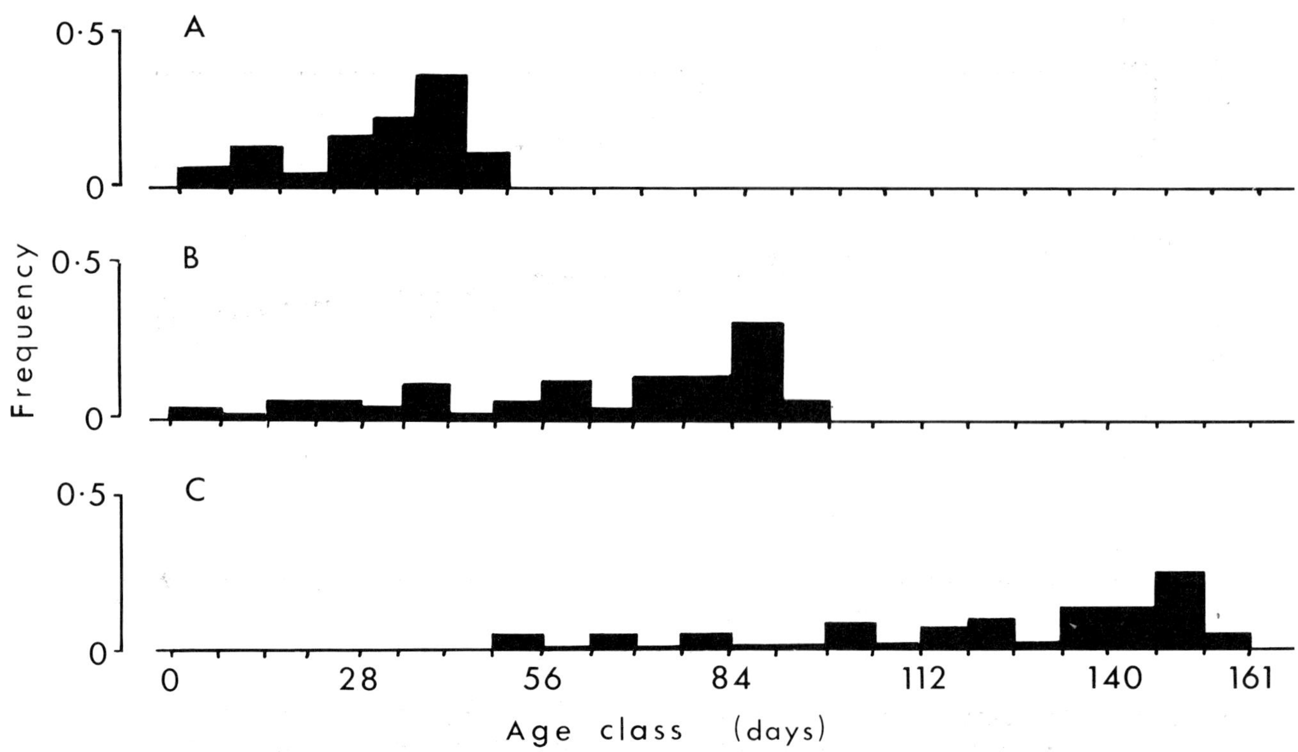


*Figure 8* **Origanum vulgare**: *Seedling age-class distribution (as a proportion of seedlings present) at three dates.*

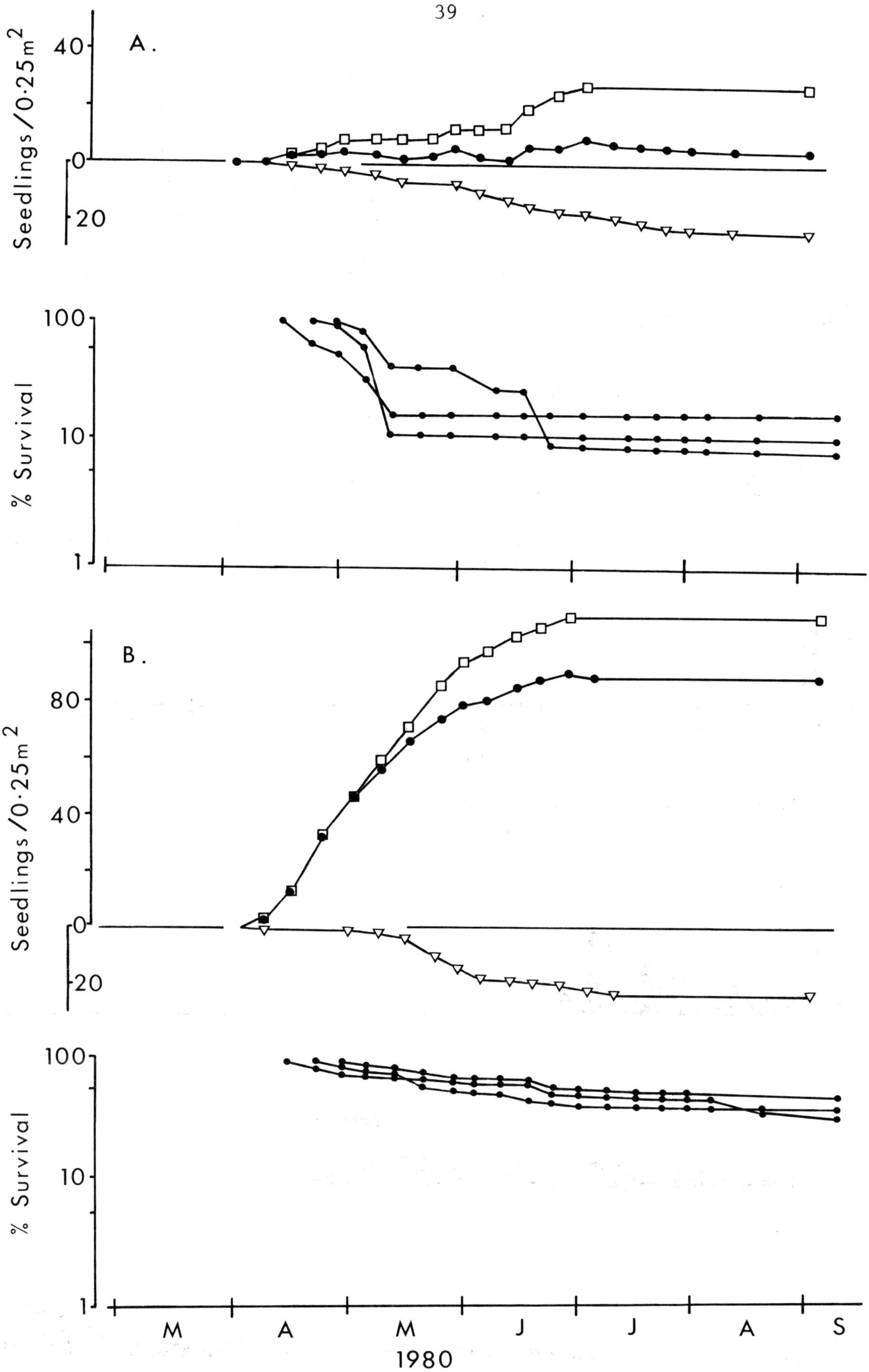


*Figure 9* **Origanum vulgare:** *Population flux diagrams and cohort survivorship curves (mean results from three replicates)*
*A: non-mulched plots* *B: mulched plots*

REFERENCES

**BASKINS, J.M. & BASKINS, C.C.** 1972. Influence of germination date on survival and seed production in a natural population of *Laevenworthia stylosa*. *Am. Midl. Nat*, 88, 318-333.

**GROSS, K.L.** 1981. Predictions of fate from rosette size in four 'biennial' plant species: *Verbascum thapsus*, *Oenothera biennis*, *Daucus carota* and *Tragopogon dubius*. *Oecologia*, 48, 209-213.

**KLEMOV, K.M. & RAYNAL, D.J.** 1981. Population ecology of *Melilotus alba* in a limestone quarry. *J. Ecol.*, 69, 33-44.

**RAYNAL, D.J.** 1979. Population ecology of *Hieracium floretinum* (Compositae) in a central New York limestone quarry. *J. appl. Ecol*, 16, 287-298.

**SKALLER, P.** 1977. *Plant colonisation and soil development in the Jamesville Quarry*. PhD thesis, State University, Syracuse, New York.

**WERNER, P.A.** 1975. Predictions of fate from rosette size in Teasel (*Dipsacus fullonum* L.). *Oecologia*, 20, 197-201.

# THE DYNAMICS OF CHALK QUARRY VEGETATION

**BRYAN G FINEGAN AND H J HARVEY**
*Department of Applied Biology, University of Cambridge*

## INTRODUCTION

Abandoned chalk quarries may become island refugia for plant species which are becoming increasingly local or rare due to the destruction of the natural or semi-natural plant communities characteristic of the areas in which such quarries occur (Hodgson; Ranson & Doody, this symposium). In some cases a refugium may hold a sufficiently large proportion of the populations of particular species in an area for pressure to arise for the conservation of the vegetation of that quarry. The conservation of vegetation requires management, which is only possible with a knowledge of the structure and dynamics of that vegetation.

The establishment of a given species in an abandoned quarry depends upon the dispersability of the propagules of that species and the tolerance of the developing individuals to the biotic and abiotic conditions which they experience. The work of Gray (this symposium) makes clear the role of chance in colonisation. The studies of Park (this symposium) elegantly demonstrate that edaphic, biological and climatic factors affecting the early stages of the life cycle may limit the number of individuals which become established from seed on bare quarry floors. The considerable diversity in the species composition of the vegetation which may be observed both within and, particularly, between abandoned quarries could arise from many factors, for example, chance, heterogeneity of the physical and chemical environment, the time since abandonment, or differences in quarry use both before and after abandonment.

If vegetation has occupied an abandoned quarry for many years then the means by which the population of a given species is maintained, and the factors which limit the size of that population, cannot be understood solely through the study of the germination of seeds and the establishment of juvenile plants. The demands made by a plant on the available resources and its sensitivity to extreme conditions will vary with ontogenetic stage, while all stages of the life cycle have to be completed before potential descendants are produced. Ideally therefore the whole life cycle of each species should be studied, together with the interactions of individuals with each other and with their environment. Such information on the structure and dynamics of plant communities is vital both for the formulation of sound management plans and as a basis for attempts to recreate communities in new locations (Down, this symposium). A study with the aim of producing such information is in progress at two adjacent chalk quarries in southern England. This paper reports some initial results.

## METHODS

The two quarries were surveyed in August 1979 and the major vegetation types were identified. During the winter of 1979-80 soil samples were taken from 100m$^2$ plots established in each vegetation type and were analysed for selected physical and chemical parameters. Eight sites, four in each quarry representing the major vegetation types, were selected for the collection of demographic data and three permanent quadrats, each 0.25m$^2$, were established in each. All plants present in these plots were mapped by pantograph at intervals of three or four weeks between May and November 1980. Particular attention was paid to part of one quarry, to which SSSI status has been assigned. No proper study of seedling recuitment was made during 1980 and new genets were not mapped until they had developed three of four leaves.

## RESULTS AND DISCUSSION

Only data collected in the SSSI are presented, a species list for this area is given in Table 12. The vegetation is dominated by a *Betula/Salix* scrub of variable density. Most of the perennial herbaceous species occur

**TABLE 12 Species of vascular plant recorded to date in this study on the SSSI (Nomenclature after Clapham *et al* 1962).**

*Aceras anthropophorum*
*Agrostis stolonifera*
*Anacamptis pyramidalis*
*Bellis perennis*
*Betula pendula*
*Blackstonia perfoliata*
*Brachypodium sylvaticum*
*Briza media*
*Carex flacca*
*Carpinus betulus*
*Centaurium erythraea*
*Clinopodium vulgare*
*Crataegus monogyna*
*Dactylis glomerata*
*Dactylorhiza fuchsii*
*Daucus carota*
*Erigeron acer*
*Festuca ovina/rubra*
*Fragaria vesca*
*Gentianella amarella*
*Hedera helix*
*Hieracium diaphanum*
*Hieracium perpropinquum*
*Hieracium pilosella*
*Inula conyza*
*Leontodon hispidus*
*Listera ovata*
*Lotus corniculatus*
*Neottia nidus-avis*
*Odontites verna*
*Ophioglossum vulgatum*
*Origanum vulgare*
*Pastinaca sativa*
*Prunella vulgaris*
*Pyrola rotundifolia*
*Quercus robur*
*Rosa* sp.
*Salix* spp.
*Thelycrania sanguinea*

throughout the area but four major communities may be recognised on the basis of changes in the total cover of vegetation and the relative abundances of species. The chief features of these communities are:

1. open scrub with sparse herbaceous vegetation, monocarpic species frequent;
2. open scrub with forb dominated turf in which *Carex flacca* is abundant;
3. denser scrub with *Brachypodium sylvaticum* and bryophytes;
4. dense scrub with deep shade and sparse ground flora.

The results of the analysis of soils from the communities are shown in Table 13.

**TABLE 13 Characteristics of the soils under the main types of vegetation within the SSSI. Values are the means (±S.E.) of five analyses, each based on two samples from randomly positioned plots of $0.25m^2$.**

| Vegetation type | pH | Nitrate $\{NO_3^-\}$ mg/ℓ | Phosphate $\{PO_4^{---}\}$ mg/ℓ |
|---|---|---|---|
| A and B | 7.82±0.02 | 2.43±0.28 | 7.28±1.26 |
| C | 7.75±0.06 | 3.11±0.81 | 10.04±0.92 |
| D | 7.76±0.02 | 4.42±0.50 | 15.28±0.72 |

Community C is most widespread. No results from community D are presented here as the species involved do not occur widely in the other three communities.

Demographic data for *Carex flacca, Hieracium pilosella* and *Leontodon hispidus* are given in Figure 10. The major features of the data appear to be similar for each vegetation type and representative examples only are given.

*Demography of* **Carex flacca**

The fates of 1,500 aerial shoots of this species were determined during 1980. *Carex* spreads clonally as a consequence of a sympodial rhizome system, each module terminating in a leafy aerial shoot, at the base of which two or more axillary buds may develop into new modules. The data therefore represent the production of modular units, though whether the death of an aerial shoot marks the death of the whole module is less clear. There is little evidence of seasonal trends in recruitment or mortality over the period of study (Figure 10A), although a summer peak in leaf biomass (as estimated by percentage cover) has been recorded in chalk downland populations (J. Mitchley, pers. comm.). Modules produce terminal inflorescences in May and June and are monocarpic. Less than 1% of the monitored modules produced seed in 1980 and no new genets were recorded. It seems that this species may be entirely maintained by clonal growth.

*Demography of* **Hieracium pilosella**

This species is a scapigerous stoloniferous rosette herb (Clapham *et al*, 1962); the fates of 325 individuals were determined during 1980.

Flowering began in May and continued to September but less than 7% of the monocarpic rosettes produced mature capitula; the development of many inflorescences ceased at an early stage and others were eaten. The production of axillary stolons, bearing terminal rosettes, which emerge at the same time as inflorescences appears to be linked to the onset of flowering (Bishop *et al*, 1978). These stolons continue to grow whatever the fate of the capitulum. Bishop *et al* (1978), working in the Breckland of East Anglia, recorded daughter rosettes as recruits from June onwards and their data show a June-July peak in recruitment. In none of the populations they studied did they observe the recruitment of new genets. In this study daughter rosettes were not treated as recruits until they became rooted in August and September. Many daughter rosettes senesced before they rooted and too few became established to compensate for the death of parent rosettes. The population was only maintained by the recruitment of new genets from seed and the greater proportion of recruits in 1980 were from this source. The mortality of these recruits was higher than that of rosettes present at the start of the study. As a consequence of incorporating new genets into the population only at the four leaf stage, and daughter rosettes only when they had rooted, the picture which emerges from this study of the within-season dynamics of *H. pilosella* differs from that observed by Bishop *et al* (1978). In this study the size of the population was greatest in July and August and numbers had fallen by October (Figure 10B).

It is clear that the populations of *H. pilosella* investigated in this study would decline in the absence of establishment from seed. Watt (1962) similarly found that Breckland populations of *H. pilosella* which established from seed in favourable years subsequently declined if no further recruitment of genets occurred. *H. pilosella* may differ in this respect from *H. florentinum*, the population of which in one abandoned limestone quarry in North America was maintained primarily by clonal growth (Raynal, 1979).

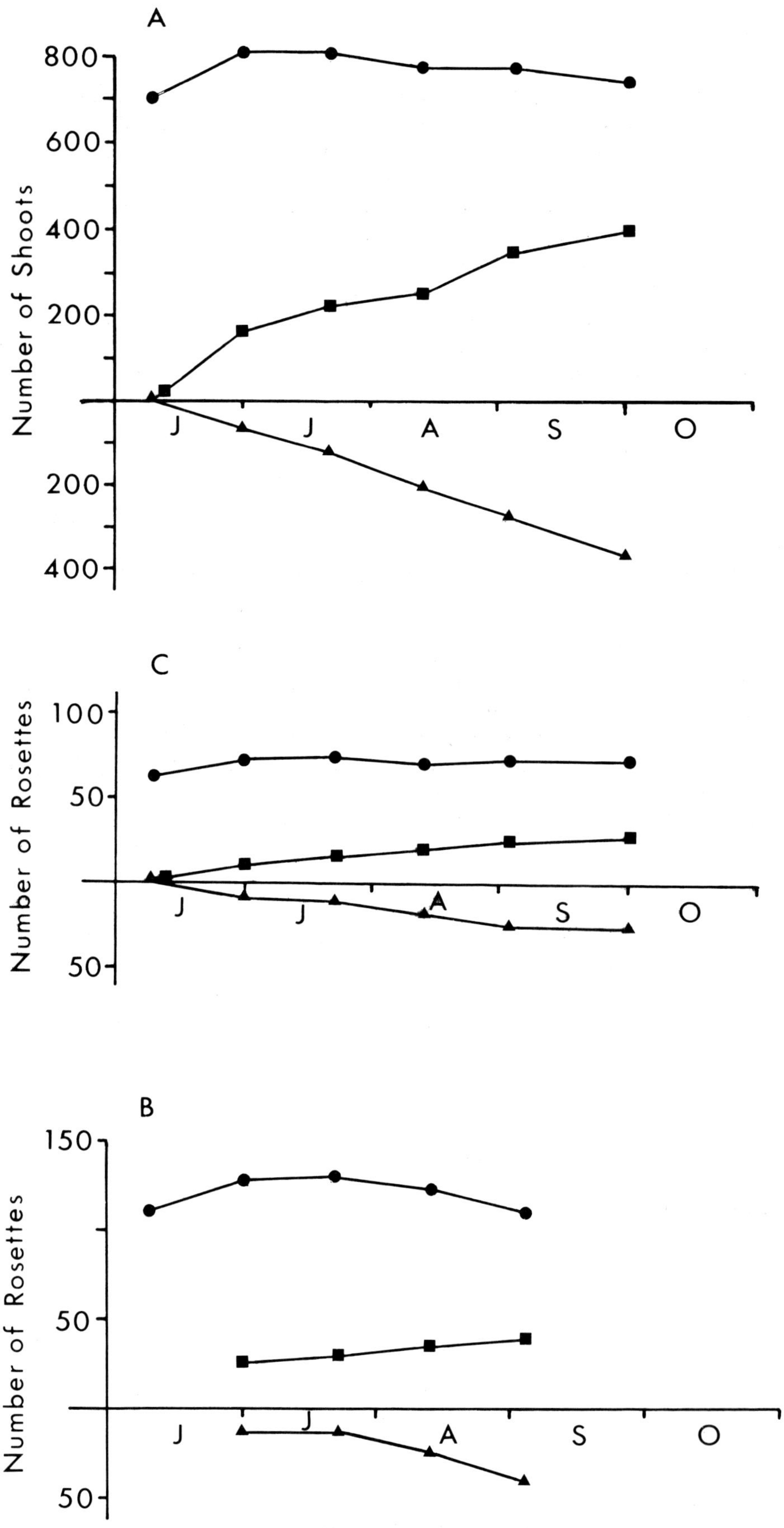


*Figure 10* *The flux of "individuals" in populations of three species in a chalk quarry in 1980. Mean numbers in 0.75* $m^2$.
*A.* **Carex flacca** *B.* **Hieracium pilosella** *C.* **Leontodon hispidus**

■—■ *cumulative births,* ▲—▲ *cumulative deaths,*
●—● *size of population*

*Demography of* **Leontodon hispidus**

This species is a scapigerous polycarpic rosette herb; one or more leaf rosettes may be produced from the same rootstock (Clapham *et al*, 1962), although in the present study it was found that only one rosette persisted, the older usually senescing. This mortality may account for up to 30% of the recorded flux of rosettes at any one site, the actual flux of genets is therefore lower than the data in Figure 10C would seem to indicate. Less than 5% of the individuals established in May 1980 produced inflorescences during the year and, as in *Hieracium pilosella*, predation of the capitula was high. Small numbers of genets became established between June and September. This low level of recruitment, combined with the apparent longevity of the polycarpic rosettes, would seem to be adequate to maintain the population.

CONCLUSIONS

It is clear from this study that, even in vegetation composed of only a few species, the phytological processes which underlie the structure of a plant community may be very diverse. A similar conclusion can be reached on the basis of studies of sympatric species of *Ranunculus* in grassland (Sarukhan, 1974) or of rare species in Teesdale (Bradshaw & Doody, 1978). Moreover the processes operating in other, superficially similar, sites may be very different, as shown by the marked contrast in the flux of individuals through the populations studied here and those investigated by Park (this symposium). If each population of every species may be maintained and regulated in different ways, then a series of unique studies may be necessary to provide the information required for sound managment. Such studies are possible and can be extremely revealing (eg Harvey and Meredith, 1981; Bradshaw & Doody 1978).

To complement demographic studies, experiment is needed to determine the extent to which individuals of different species interfere with each other. Also important in the abandoned quarry situation, especially if recreation of vegetation is the aim, is investigation of the role of abiotic factors in determining the abundance and performance of species. Such studies are in progress.

ACKNOWLEDGEMENTS

This work was funded by a Natural Environment Research Council C.A.S.E. award to B.G.F. in which Thurrock Borough Council is the collaborating body providing funds and facilities. Mr R N Humphries gave valuable help in the planning of the research programme.

REFERENCES

**BISHOP, G.G., DAVY, A.J. & JEFFERIES, R.L.** 1978. Demography of *Hieracium pilosella* in a Breck grassland. *J. Ecol.*, 66, 615-629.

**BRADSHAW, M.E. & DOODY, J.P.** 1978. Plant population studies and their relevance to nature conservation. *Biol. Conserv.*, 14, 223-242.

**CLAPHAM, A.R., TUTIN, T.G. & WARBURG, E.F.** 1962. *Flora of the British Isles.* Cambridge University Press.

**HARVEY, H.J. & MEREDITH, R.C.** 1981. Ecological studies on *Peucedanum palustre* and their implications for conservation management at Wicken Fen, Cambridgeshire. In: *The biological aspects of rare plant conservation,* edited by H. Synge, 365-378. Chichester: John Wiley.

**RAYNAL, D.J.** 1979. Population ecology of *Hieracium florentinum* (Compositae) in a central New York limestone quarry. *J. appl. Ecol.*, 16, 287-298.

**SARUKHAN, J.** 1974. Studies on plant demography: *Ranunculus repens* L., *R. bulbosus* L. and *R. acris* L. II. Reproductive strategies and seed population dynamics. *J. Ecol.*, 62, 152-177.

**WATT, A.S.** 1962. The effect of excluding rabbits from grassland A (Xerobrometum) in Breckland, 1936-60. *J. Ecol.*, 50, 181-198.

# SUCCESSION

**A D BRADSHAW, R H MARRS* AND R D ROBERTS†**
*Department of Botany, University of Liverpool.*

It is not much of generalisation to say that the soils to be found in quarries are skeletal; there is plenty of bone but no flesh. They are not really even entisols in the American terminology, they are so juvenile and deficient. From the point of view of plant growth, the most important deficiency is nitrogen, for it is not a component of rock materials and it has not had time to be fixed by living organisms and accumulated in organic matter, then to be available by mineralisation. The second major deficiency is available phosphorus, because if it is present in rocks it will be locked up, and it, too, will not have been accumulated in organic matter.

Plants can arrive at a site and germinate without nutrients, but they cannot grow. So plant size and the development of the biomass of the ecosystem is dependent on nutrient supply. There can be complex interactions; for example, lack of nutrients will lead to an incomplete vegetation cover and possible erosion of seedlings that have germinated, and to lack of an adequate root system with death of seedlings from drought. Until nutrients, particularly nitrogen, are available, the whole ecosystem is in jeopardy, even if the species are tolerant of low nutrient supply. Where will these nutrients come from? The nitrogen can only come from biological fixation or in small amounts (about 10 kg/ha/yr) in precipitation. Very little phosphorus comes in precipitation (less than 1 kg/ha/yr), so accumulation can only occur through the weathering of rocks. The other nutrients will come from precipitation and from rocks, and are not usually a problem because they are released more readily than phosphorus (Russell 1973).

It is easy to get species to grow in rock waste or subsoil if nutrients, mainly N and P, are provided. But almost always growth soon stops. In nearly all the situations that we have analysed this is due to lack of nitrogen; if nitrogen alone is added or in combination with other elements, growth recommences immediately (Bradshaw *et al* 1975; Bloomfield *et al* 1982). It then ceases again unless more nitrogen is added. There seem to be three reasons for this. Firstly, if nitrogen is applied to skeletal soils without vegetation, most of it is leached. Secondly, more nitrogen is required by plants than any other nutrient. Thirdly, and crucially, nitrogen is held in soils almost entirely in organic matter and released by mineralisation; adequate amounts of nitrogen are therefore only released if there is a large store of nitrogen in organic matter, because the rate of mineralisation is only about 2%/yr (Reuss & Innis 1977). In a temperate climate a rather unproductive ecosystem might have an annual productivity of 5,000 kg/ha containing 1% N and therefore an annual requirement of nitrogen of 50 kg/ha. To provide this, if we presume an annual supply of 10 kg/ha in precipitation and a mineralisation rate of 2%/yr, a nitrogen store of 2000 kg/ha in a mineralisable form in the soil would be necessary. The soil store could be less if there was more rapid cycling in some parts of the decomposer pathway than others, eg in litter. But whatever happens there is need for quite a substantial capital.

We have derived an estimate of the minimum nitrogen capital required, by an examination of the nutrients contained in different natural ecosystems

* present address: *Institute of Terrestrial Ecology, Monks Wood Experimental Station, Huntingdon.*
† present address: *Department of Biology, University of Essex*

developed on china clay wastes (Dancer *et al* 1981; Marrs *et al* 1981; Roberts *et al* 1981). There is a series which is clearly a developmental sequence. All nutrients accumulate to a certain extent, but nitrogen is the one that shows the greatest accumulation (Table 14). Because of the previous arguments, it is clear that nitrogen is the most critical in determining the succession. We can therefore use the values as indicators of the minimum nitrogen levels necessary for self-sustaining scrub, shrub and woodland vegetation, because when species from the latter stages of the succession are found as colonists in the earlier stages they do not grow or grow only very slowly. Values for mineralisable nitrogen correlate with these totals (Skeffington & Bradshaw 1981). It is interesting that the first substantial community, of *Salix*, appears to have a minimal soil nitrogen requirement of only 700 kg/ha. This is lower than our theoretical calculation, perhaps because the overall rate of mineralisation is more than 2%/yr, or because productivity is lower.

The major source of build up of the nitrogen store is by legume, mainly *Ulex* sp; little nitrogen fixation takes place elsewhere (Skeffington & Bradshaw 1980). Indeed the whole succession appears to be controlled by the appearance of *Ulex*; if it does not appear, a very poor heath vegetation persists for many years. The crucial control by legume is well illustrated by the invasion of a very successful alien legume, *Lupinus arboreus*, which can accumulate about 180 kg N/ha/yr. This will invade mica lagoons when the only other vegetation is a scattering of very depauperate plants of *Holcus lanatus*. After the *Lupinus* has been growing for three years each individual plant becomes surrounded by a bright green ring of grass; when the *Lupinus* dies after six years it leaves behind a substantial sward of *Holcus* in which *Salix* flourishes (Palaniappan *et al* 1979) (Figure 11).

**TABLE 14 Dominant species, biomass and nutrient content of the main ecosystem types on naturally-colonised china clay waste (kg/ha); there is a progressive build up of nitrogen.**

| Ecosystem type | Dominant species | Age (yrs) | Biomass | N | P | K | Ca | Mg | |
|---|---|---|---|---|---|---|---|---|---|
| Primary | *Lupinus arboreus* | 17 | 10888 | 110 | 45 | 103 | 46 | 19 | (plant) |
| | | | | 181 | 107 | 614 | 99 | 38 | (soil) |
| | *Calluna vulgaris* | 50 | 21445 | 151 | 15 | 84 | 29 | 20 | (plant) |
| | *Ulex europaeus* | | | 672 | 152 | 1478 | 66 | 94 | (soil) |
| | *Sarothamnus scoparius* | | | | | | | | |
| Intermediate | *Salix atrocinerea* | 53 | 30292 | 259 | 11 | 344 | 56 | 31 | (plant) |
| | | | | 722 | 119 | 1439 | 75 | 83 | (soil) |
| Mature | *Rhododendron ponticum* | 92 | 157182 | 581 | 71 | 358 | 606 | 122 | (plant) |
| | *Betula pendula* | | | 1189 | 126 | 704 | 155 | 130 | (soil) |
| | *Quercus robur* | | | | | | | | |

However, other nutrients must not be forgotten. In china clay wastes almost every major plant nutrient other than potassium is extremely deficient, and must be added if normal grassland including agricultural legumes, which are extremely powerful nitrogen fixers, are to be established. *Ulex europaeus* and *Lupinus arboreus* appear perhaps to be exceptional. But, although they can tolerate low nutrient levels, they will not flourish on raw china waste unless some calcium and phosphorus has been added, about 500 kg/ha of $CaCO_3$ and 25 kg/ha of P. *Lupinus arboreus*, in fact, has a higher requirement than *Ulex europaeus*. Other legumes normally associated with poor soils, such as *Anthyllis vulneraria* and *Lotus corniculatus* have distinct Ca and P requirements also (Figure 12).

*Figure 11 A group of bushes of* **Lupinus arboreus** *growing on mica waste showing the development of a grass sward once the bushes are more than 3 years old.*

It follows, therefore, that nitrogen accumulation and hence vegetation development in quarry ecosystems depends on calcium and phosphorus availability. This is not a particularly profound conclusion until one realises that there are many quarry situations where calcium and phosphorus are limiting and as a result ecosystem development is very slow. Acidic rocks are clearly deficient in calcium and phosphorus and their release is restricted by the hardness of the rock. But it must be remembered that these rocks cannot only differ in their hardness but also in the calcium and phosphorus that they contain (Clarke 1924) (Table 15). As a result, there are considerable differences in the rate at which ecosystems develop upon them, as any contrast such as between natural rhyolite and basalt scree slopes shows.

On alkaline rocks there is obviously sufficient calcium and often appreciable amounts of total phosphorus, but most of this is unavailable to plants in the form of apatite ($Ca_2$ $(PO_4)_3$). Further inputs of phosphorus from elsewhere will be complexed also. A major phosphorus requirement can easily be demonstrated (Bradshaw *et al* 1977). Since phosphorus inputs from the atmosphere are very low, ecosystem development will be very slow indeed, and there will surely be a long period before phosphorus accumulation reaches a threshold of availability, allowing legumes to prosper. Coupled with the fact that there are large numbers of species which are adapted to calcareous habitats, this could explain why calcareous quarries and wastes, such as

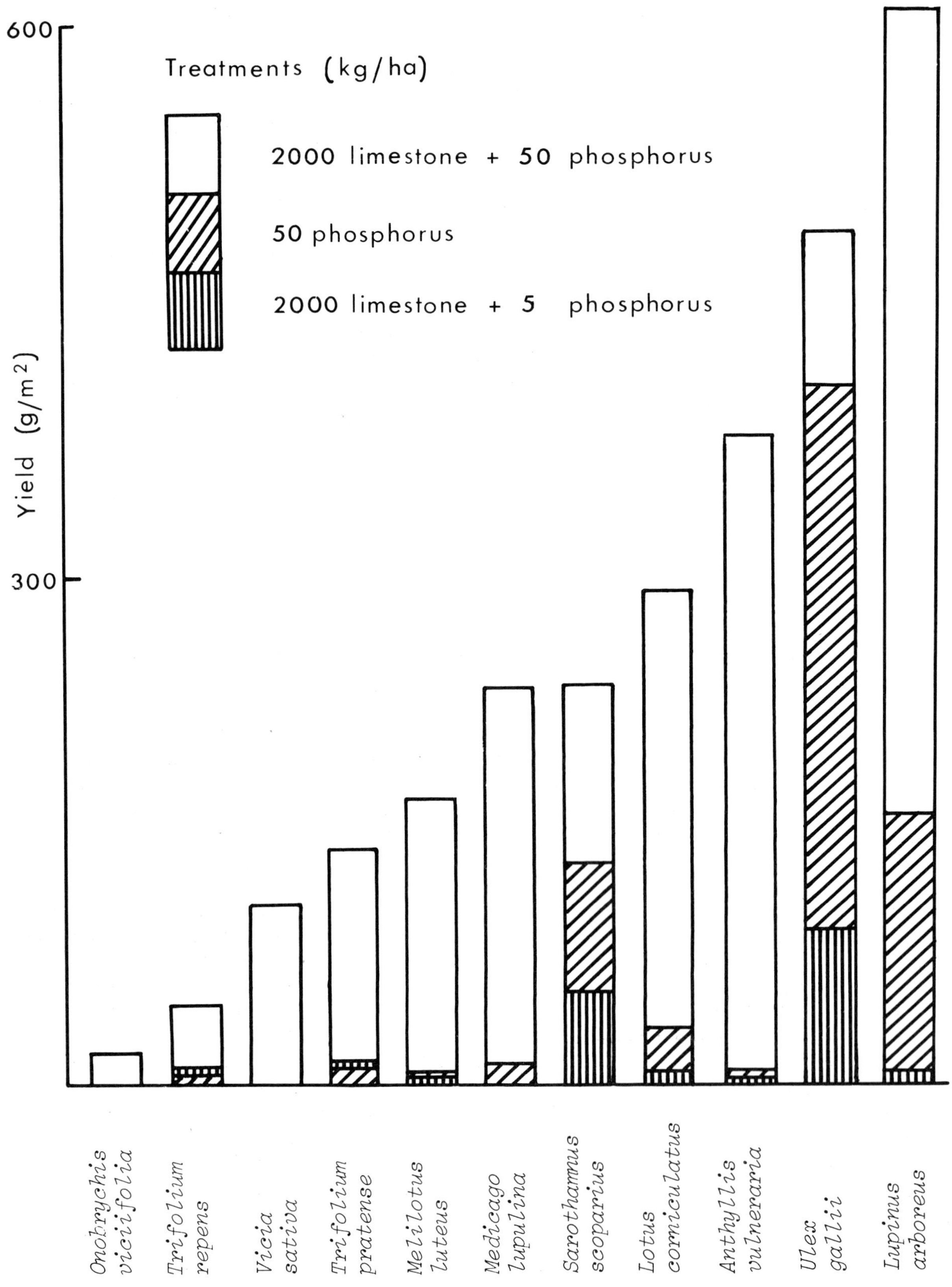


*Figure 12* *The growth of various legume species on sand waste without and with the addition of small amounts of lime and phosphorus: the growth of all species is improved by some additions.*

TABLE 15 The average chemical composition (%) of different types of rock: there is considerable variation in critical nutrients such as calcium and phosphorus

| | rhyolites | granites | diorites | basalts | dolerites | shales | sand-stones | lime-stones |
|---|---|---|---|---|---|---|---|---|
| $SiO_2$ | 72.80 | 70.18 | 58.90 | 49.06 | 50.48 | 58.10 | 78.33 | 5.19 |
| $TiO_2$ | 0.33 | 0.39 | 0.76 | 1.36 | 1.45 | 0.65 | 0.25 | 0.06 |
| $Al_2O_3$ | 13.49 | 14.47 | 16.47 | 15.70 | 15.34 | 15.40 | 4.77 | 0.81 |
| $Fe_2O_3$ | 1.45 | 1.57 | 2.89 | 5.38 | 3.84 | 4.02 | 1.07 | 0.54 |
| FeO | 0.88 | 1.78 | 4.04 | 6.37 | 7.78 | 2.45 | 0.30 | - |
| MnO | 0.08 | 0.12 | 0.12 | 0.31 | 0.20 | - | - | - |
| MgO | 0.38 | 0.88 | 3.57 | 6.17 | 5.79 | 2.44 | 1.16 | 7.89 |
| CaO | 1.20 | 1.99 | 6.14 | 8.95 | 8.94 | 3.11 | 5.50 | 42.57 |
| $Na_2O$ | 3.38 | 3.48 | 3.46 | 3.11 | 3.07 | 1.30 | 0.45 | 0.05 |
| $K_2O$ | 4.46 | 4.11 | 2.11 | 1.52 | 0.97 | 3.24 | 1.31 | 0.33 |
| $H_2O$ | 1.47 | 0.84 | 1.27 | 1.62 | 1.89 | 5.00 | 1.63 | 0.77 |
| $P_2O_5$ | 0.08 | 0.19 | 0.27 | 0.45 | 0.25 | 0.17 | 0.08 | 0.04 |
| $CO_2$ | - | - | - | - | - | 2.63 | 5.03 | 41.54 |
| $SO_3$ | - | - | - | - | - | 0.64 | 0.07 | 0.05 |
| BaO | - | - | - | - | - | 0.05 | 0.05 | - |
| C | - | - | - | - | - | 0.80 | - | - |

those produced by the Leblanc process described by Miss Gray, maintain open habitats and species diversity for so long.

From the point of view of quarries as sites of high species-diversity and refuges for uncommon species, a very slow rate of ecosystem development and succession is a good thing. In our opinion the key to this slow rate lies in the rate at which leguminous and other nitrogen fixing plants develop. This can be affected by many factors, of which calcium and phosphorus are particularly important.

But even early colonists require some nitrogen and other basic nutrients for growth, so the positive creation of areas of nature conservation interest in quarries (Bradshaw 1977) can depend on a small, carefully controlled input of missing nutrients, particularly phosphorus. But this must not be overdone or the sites could become too eutrophic and the desired species disappear under an effusion of more vigorous species. However, the nature of the quarry habitat makes it very unlikely that all areas of a quarry will become completely eutrophic, even with rather uncontrollable application of nutrients, and it is unlikely that the general public will be very inspired by areas, created to be nature reserves, which are bare rock for their first fifty years. So, even for purposes of nature conservation, some input of nutrients seems essential. If the need is for establishment of a robust vegetation cover for screening or engineering purposes, whether it is of grass, shrubs or trees, the crucial importance of ecological engineering to ensure the rapid accumulation of an adequate nitrogen capital cannot be over-emphasised.

REFERENCES

**BLOOMFIELD, H., HANDLEY, J.F. & BRADSHAW, A.D.** 1982. Nutrient deficiencies and the aftercare needs of reclaimed derelict land. *J. appl. Ecol.* (in press).

**BRADSHAW, A.D.** 1977. Conservation problems in the future. *Proc. R. Soc. B,* 197, 77-96.

**BRADSHAW, A.D., DANCER, W.S., HANDLEY, J.F. & SHELDON, J.C.** 1975. The biology of land revegetation and the reclamation of the china clay wastes of Cornwall. In: *The ecology of resource degradation and renewal* edited by M.J. Chadwick and G.T. Goodman, 363-384. Oxford : Blackwells. (Symp. Br. Ecol. Soc. no. 15)

**BRADSHAW, A.D., HUMPHRIES, R.N., JOHNSON, M.S. & ROBERTS, R.D.** 1977. The restoration of vegetation on derelict land produced by industrial activity. In: *The breakdown and restoration of ecosystems,* edited by M.W. Holdgate and M.J. Woodman, 249-278. New York : Plenum. (NATO Conference Series: series 1: Ecology, v.3)

**CLARKE, F.W.** 1924. The data of geochemistry. 5th ed. *Bull. U.S. geol. Surv.* no. 770.

**DANCER, W.S., HANDLEY, J.F. & BRADSHAW, A.D.** 1977. Nitrogen accumulation in kaolin mining wastes in Cornwall. I. Natural communities. *Pl. Soil,* 48, 153-167.

**MARRS, R.H., ROBERTS, R.D., SKEFFINGTON, R.A. & BRADSHAW, A.D.** 1981. Ecosystem development on naturally-colonized china clay wastes. II. Nutrient compartmentation. *J. Ecol.,* 69, 163-169.

**PALANIAPPAN, V.M., MARRS, R.H. & BRADSHAW, A.D.** 1979. The effect of *Lupinus arboreus* on the nitrogen status of china clay wastes. *J. appl. Ecol.,* 16, 825-831.

**REUSS, J.O. & INNIS, G.S.** 1977. A grassland nitrogen flow simulation model. *Ecology,* 58, 379-388.

**ROBERTS, R.D., MARRS, R.H., SKEFFINGTON, R.A. & BRADSHAW, A.D.** 1981. Ecosystem development on naturally-colonized china clay wastes. I. Vegetation changes and overall accumulation of organic matter and nutrients. *J. Ecol.,* 69, 153-161.

**RUSSELL, E.W.** 1973. *Soil conditions and plant growth.* 10th ed. London: Longmans.

**SKEFFINGTON, R.A. & BRADSHAW, A.D.** 1980. Nitrogen fixation by plants grown on reclaimed china clay wastes. *J. appl. Ecol.,* 17, 469-477.

**SKEFFINGTON, R.A. & BRADSHAW, A.D.** 1981. Nitrogen accumulation in kaolin mining wastes in Cornwall. IV. Sward quality and the development of a nitrogen cycle. *Pl. Soil,* 62, 439-451.

# DISCUSSION

**CHAIRMAN: H J HARVEY**
*Department of Applied Biology, University of Cambridge*

Two major topics dominated this discussion - the role of woody species as primary colonisers and the importance of the accumulation of reserves of mineral nutrients in the development of vegetation.

Dr P G Grubb (Department of Botany, Cambridge) considered that the early invasion of woody species onto bare soil was not unusual and noted that in the tropics trees had long been recognised as primary colonisers. He quoted as examples *Nothofagus* colonising land slips in South America, *Pinus canariensis* colonising volcanic ash in the Canaries and *Buddleja* in Asia. Other examples of the development of woody vegetation on abandoned sites, without a preceding herbaceous vegetation, were quoted and this would appear to be a common phenomenon. It is interesting to note that studies carried out in Britain in the 1930s showed woody species to be important colonisers of bare chalk in abandoned quarries (Hope-Simpson 1940; Locket 1945). The first of these studies emphasised the importance of grazing, by rabbits, in determining the success of such colonisations. This lends support to the observations of Park that animals may have an important impact on the development of vegetation.

Dr Grubb also questioned the generality of the view that vegetation succession was dependent upon the mineral nutrient status of the system. He observed that some species of plant (eg *Eucalyptus obliqua)* appeared to be capable of developing a large biomass in systems low in mineral nutrients, the mineral content of the certain tissues (eg heart-wood) being very low (eg Attiwell 1980). Detailed analyses of plant tissues are needed to discover whether or not the biological pool of mineral nutrients reflected levels in the soil. In some situations an increase in the mineral nutrient reserves of a soil could be a consequence, rather than a cause, of vegetation development. For example, the invasion of hawthorn *Crataegus monogyna* scrub into chalk grassland can increase the concentration of mineral nutrients, especially phosphate, in the soil, probably by the deposition in litter of nutrients extracted from deeper horizons. Clearly such a mechanism might not operate where there are no nutrient reserves at depth, as in the china clay wastes studied by Professor Bradshaw and his colleagues. There was also discussion of the magnitude of the input of mineral nutrients in rainfall, and whether or not this was adequate to account for the amounts accumulated in systems. The possible importance of root/shoot ratio s in determining success in colonisation of bare sites was mentioned. It was suggested that some of the ecotypes of trees which were most successful in colonising coal spoil have high root/shoot ratios.

The validity and value of island biogeography models in studies of colonisation were discussed. Several examples were noted where the relationship between the size of the study area and the number of species of plant recorded did not fit theoretical predictions. Reference was made to recent papers (Gilbert 1980) which questioned the assumptions and meaning of the current models.

REFERENCES

**ATTIWELL, P.M.** 1980. Nutrient cycling in a *Eucalyptus obliqua* forest. IV Nutrient uptake and nutrient return. *Aust. J. Bot.*, 28, 199-222.
**GILBERT, F.S.** 1980. The equilibrium theory of island biogeography: fact or fiction? *J. Biogeogr.*, 7, 209-235.
**HOPE-SIMPSON, J.F.** 1940. Studies of the vegetation of the English Chalk. VI Late stages in succession leading to chalk grassland. *J. Ecol.*, 28, 386-402.
**LOCKET, G.H.** 1945. Observations on the colonisation of bare chalk. *J. Ecol.*, 33, 205-209.

# RESTORATION ECOLOGY

## THE ESTABLISHMENT OF VEGETATION ON QUARRY MATERIALS: PHYSICAL AND CHEMICAL CONSTRAINTS

**R N HUMPHRIES**
*Department of Applied Biology, University of Cambridge*

### INTRODUCTION

The chance of a plant establishing from seed on both bare and partially colonised quarry wastes and surfaces is low, owing to the physical and chemical constraints imposed by the materials and their microclimate (Humphries 1977a; Skaller 1977; Klemow and Raynal 1981; Park 1981). This is reflected in the time taken for the development of a complete vegetation cover by natural colonisation, and the necessity of amelioration treatment for the successful establishment of a vegetation cover in reclamation schemes (Humphries 1977a; 1977b).

The major constraints on establishment from seed are the supply of water and mineral nutrients. The germination of seeds, provided they are not in a dormant condition and are viable, is largely determined by water uptake and hence the effect of water supply. Other factors, like high temperatures (including secondary dormancy) and the loss of seed (burial, removal), may also affect the numbers of seedlings recruited. The survival of the seedlings is dependent on the supply of water and mineral nutrients. Again, other factors, like defoliation by animals or prevention of root growth by cementation, may affect the number of survivors. The occurrence and intensity of the major constraints is extremely variable, both between and within quarries and both between and within years, owing to differences in quarry materials and weather conditions (Humphries 1980; Park 1981).

The nature of the quarry materials on many sites determines the supply of water for germination (through surface texture) and for seedling growth (Humphries 1976; 1977a). The supply of mineral nutrients for seedling growth is largely determined by particle size distribution and clay mineral content (on raw uncolonised material) and organic matter (when vegetated) (Humphries 1977a, 1977b). The importance of surface texture, water and mineral nutrients for seedling recruitment and survival is illustrated by reference to two simple greenhouse experiments with grassland species.

### EFFECTS OF SURFACE TEXTURE, WATER AND NUTRIENTS ON *MEDICAGO LUPULINA*

In the first experiment, a contrast in surface texture (smooth and rough) was achieved by using two grades of crushed limestone (sand and gravel). Three levels of water supply were created by the addition and mixing of either a silt-loam subsoil (giving a two-fold increase in water holding capacity over the limestone material) or milled sphagnum peat (three-fold increase). Three levels of mineral nutrient content (increases of two and three-fold over the limestone material) were created by the differential loss by leaching of NPK fertiliser applied uniformly as small dosages every five days. Least nutrients were retained in the unamended limestone treatment and most in the sphagnum treatment owing to differences in particle size distribution and cation exchange capacity.

In this way a gradient of mineral nutrient and water supply was produced. The combined supply of water and mineral nutrients is hereafter referred to as *resource level*. An agricultural variety of *Medicago lupulina* was used as a reference species; this germinated within 5 days in ideal conditions and had a viability of 98%.

**TABLE 16 The effect of surface texture on the time taken for seeds of *Medicago lupulina* to germinate.**

| Surface texture | Number of days for 90% germination |
|---|---|
| smooth | 23 |
| rough | 13 |
| smooth with mulch | 15 |

Germination of surface sown seed was appreciably quicker on the rough textured gravel than on the smooth surfaced sand (Table 16). The application of a mulch to the sand also speeded up germination (as did seed burial - data not given). Both the rough surfaces and the mulch had a better seed spoil contact for rapid water uptake. The rapid germination on the rough textured material, and under the mulch, resulted in fewer and more even-aged cohorts of seedlings than on the smooth material. An increase in resource level did not affect germination rate of surface sown seed. Surface texture did not affect the number of seedlings surviving whereas resource level did. There was an increase in survival and in the size of individual plants with an increase in resource level (Table 17 and 18).

**TABLE 17 The effect of resource level on the survival of newly recruited *Medicago lupulina* seedlings.**

| Resource level | Percentage of recruited seedlings surviving |
|---|---|
| lowest | 32 |
| intermediate | 48 |
| highest | 68 |

**TABLE 18 The effect of resource level on the size of individual seedling of *Medicago lupulina***

| Resource level | Mean dry weight (mg) of individual plants (shoots + roots) |
|---|---|
| lowest | 175 |
| intermediate | 555 |
| highest | 785 |

## EFFECTS ON OTHER GRASSLAND SPECIES

A second and similar experiment was designed to determine if the effects of surface texture and resource level were the same for other grassland species; 20 were selected from lists of those commonly associated with limestone quarry materials and floristically rich limestone grasslands.

The seed of all species (after the appropriate stratification or scarification to break dormancy) germinated more rapidly on the rough textured material. However, most species differed in their rates of germination and most took longer to germinate than *M. lupulina*. The survival of all species was increased by an increase in resource level but the responses to intermediate and high resource levels differed. *Arrhenatherum elatius* and *P. lanceolata* behaved like *M. lupulina*; the others all showed better survival with intermediate or high resource levels or both (Table 19). Similarly, an increase in resource level increased the size of individual plants for all species. However, some species were more affected and others less affected than *M. lupulina* at the intermediate and/or highest resource level (Table 20).

**TABLE 19 The effect of an increase in resource supply on the seedling survival of 20 grassland species+ relative to the response of *Medicago lupulina**.**

Species more responsive at intermediate and highest resource levels.

*Agrostis stolonifera, Briza media, Festuca ovina, Festuca rubra, Lolium perenne*, Achillea millefolium, Euphrasia officinalis, Geranium robertianum, Leontodon hispidus, Prunella vulgaris, Trifolium pratense.*

Species more responsive only at the highest resource level.

*Poa annua, Chrysanthemum leucanthemum.*

Species more responsive only at the intermediate resource level.

*Centaurea nigra, Linum catharticum, Lotus corniculatus, Poterium sanguisorba.*

Species more responsive at the intermediate but less so at the highest resource level.

*Geranium sanguineum.*

Species responding like *Medicago lupulina*.

*Arrhenatherum elatius, Plantago lanceolata.*

\+ seed collected from wild population on highly calcareous soils.
* forage varieties of commercial origin.

**TABLE 20 The effect of an increase in resource level on the size of an individual (for 20 grassland species) relative to the response of *Medicago lupulina*.**

Species more responsive at the highest resource level.

*Agrostis stolonifera, Festuca ovina, Festuca rubra, Prunella vulgaris.*

Species less responsive at the intermediate but more so at the highest resource level.

*Lolium perenne, Achillea millefolium, Centaurea nigra, Euphrasia officinalis.*

Species less responsive at the intermediate resource level.

*Arrhenatherum elatius, Briza media, Poa annua, Chrysanthemum leucanthemum, Geranium robertianum, Geranium sanguineum, Leontodon hispidus, Linum catharticum, Lotus corniculatus, Plantago lanceolata, Trifolium pratense.*

Species less responsive at the intermediate and at the highest resource levels.

*Poterium sanguisorba.*

## DISCUSSION

### *Seedling recruitment and survival*

The experiments showed that the rate of seedling recruitment (ie germination) could be affected by the surface texture of a quarry material. This was largely due to the effect on rate of water uptake (a function of contact between seed and quarry material). The differences in rates of germination between the species can be accounted for by differences in seed size and shape, seed-coat permeability and physiology. A rapid and short period of recruitment is most likely in conditions of unlimited water supply. In less favourable conditions rough textured surfaces can have similar effects. If the recruitment period then coincides with unfavourable conditions for seedling survival (ie low water and mineral nutrient supply), seedling mortality will be high. This could result in either the failure of a species to establish or in a very small population. A prolonged germination period might ensure the survival of some seedlings, if there was a fluctuation between favourable and unfavourable conditions or if there was a catastrophe (eg erosion, defoliation, frost). Conversely, rapid germination would increase the chance of survival in circumstances when the favourability of conditions are declining with time (eg during late spring and autumn). As species differed in their spread of germination - eg *Medicago lupulina* (narrow) compared with *Chrysanthemum leucanthemum* (broad) - it is likely that the importance of surface texture for their establishment will also vary. It will undoubtedly be of great importance for those species with a short germination period, like most agriculturally bred grasses and legumes, and for those which are short lived (annuals) or die after flowering, in which maintenance of the population is dependent on re-establishment from seed. Some populations of seed are a mixture of dormant and germinable seed; for those species surface texture may be of less importance.

Surface texture may also be of little significance when seed has become fully imbibed during dormancy and for species establishing as a result of clonal growth (Finegan and Harvey 1981). In these ways, surface texture may serve to selectively eliminate species and determine the initial population size of those species establishing from seed.

The experiments demonstrated that resource supply (as water and mineral nutrients) is a major factor in determining seedling survival. Low resource levels can result in small populations, even if there has been a high recruitment from seed. As species differ in their response to resource level, then the level of resource may act selectively to eliminate individual species. The level of resource supply certainly restricts the amount of plant growth on most raw and colonised quarry wastes. The level of supply is generally higher in well colonised material, but the levels may still be sub-optimal and can fluctuate considerably during the growing season.

The level of resource supply is of particular importance as it can determine the rate of plant growth. A greater chance of survival, during unfavourable conditions, seems to be ensured by the attainment of a critical size and/or developmental stage (Park 1981; Humphries and Barker unpublished). The *critical size* correlates well with the amount of root growth, which on a variety of spoil materials is determined by mineral nutrient supply. The *critical stage* for many species is between the cotyledon and first or second leaf developmental stages of the seedling, when mortality seems to be the highest if resource levels are low. Therefore it is important for the establishment of seedlings that the resource levels are sufficient during the critical period for rapid plant development.

Together, surface texture and resource level may be major determinants of vegetation establishment and composition on quarry materials, whether it is on raw or partially colonised material. It is therefore of considerable practical importance that due attention should be given to these characteristics when considering the establishment of vegetation from seed, whether it is by natural colonisation or by purposeful sowing.

*Implications for colonisation*

The process of colonisation is slow owing to the low survival of recruits from seed, even when there is an ample seed input to vegetate a site (Skaller 1977). The chance of establishment, hence rate of colonisation could be increased by raising the resource level. This is particularly easy if the limiting factor is mineral nutrient supply as fertilisers can be applied. The recruitment period can be modified by altering the effects of surface texture to suit the pattern of resource availability, for example by mulching (Plate 5). These techniques are equally applicable to raw quarry spoils and to partially colonised materials. Interference at different times of the year and at different intensities could be employed to select for groups of species and rates of vegetation development.

*Plate 5 The application of mulch (left) increased the rate of colonisation of a quarry spoil. Compare with the untreated right side.*

It may also be possible from the characterisation of the surface texture and resource supply of quarry materials, to assess the potential of quarry materials to promote colonisation. In this way materials which are favourable to colonisation may be selected for landscape and restoration work during the development of a quarry. Such an assessment may also be of use in the assessment and planning of both abandoned and active quarries for nature conservation (Humphries 1980).

*Implications for revegetation*

In most revegetation schemes rapid establishment and growth (erosion control, landscape) and/or a specific vegetation composition (amenity, agriculture, nature conservation) are required. To achieve this, attention must be given to the surface texture and resource supply characteristics of the materials when making specifications for site treatment (Plates 6 and 7). In many schemes which 'fail', the cause can be attributed to their neglect. For economic, practical and political reasons such chance establishment is not acceptable. To ensure that the objectives are achieved from the beginning, an appropriate amelioration treatment needs to be employed (Humprhies 1977a; 1977b). Some of the effects of surface texture can be overcome by seed placement (ie drilling, harrowing) instead of broadcasting, when there is site access. When access is restricted (eg quarry faces) then mulch treatments can be used to achieve similar effects (Humphries 1979). Particular attention needs to be given to resource supply; it is important to match the application of manures, fertilisers etc to the pattern(s) of seedling recruitment. Alternatively, quarry materials more conducive to plant establishment and growth could be identified before or during quarry development. These should be used specifically for the revegetation schemes. Not only are amelioration treatments and the selection of suitable quarry materials important for the establishment of individual plants and species, but they can be effectively used to increase the potential range of vegetation types that can be established on quarry materials (Humphries 1977b).

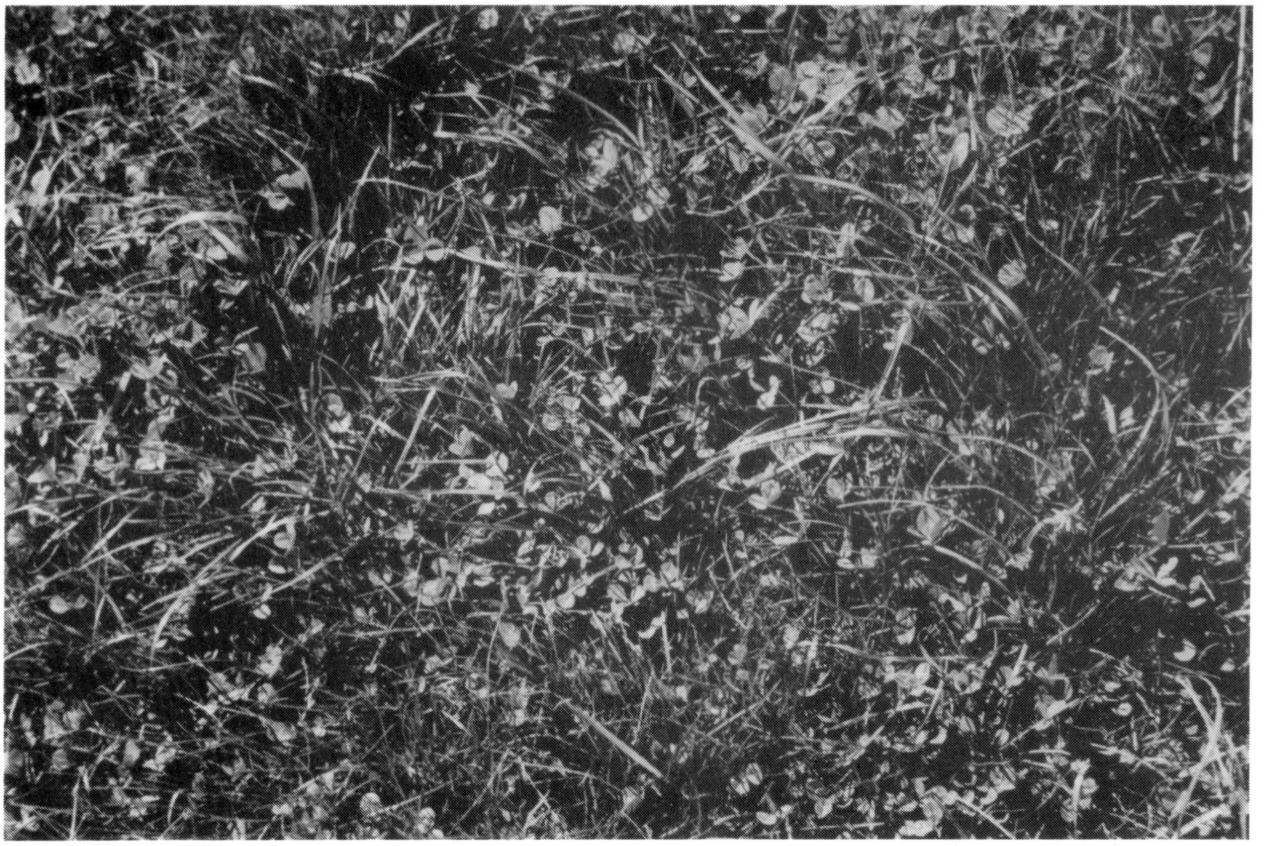

*Plate 6 A balanced grass/legume sward (for grazing) produced on crushed rock.*

*Plate 7 A species-rich chalk grassland assemblage produced on solid chalk.*

## REFERENCES


**FINEGAN, B.G. & HARVEY, H.J.** 1981. The dynamics of chalk quarry vegetation. *This Symposium.*

**HUMPHRIES, R.N.** 1976. The revegetation of limestone quarries. In: *Papers of the land reclamation conference. Grays 1976*, 443-480. Grays: Thurrock Borough Council.

**HUMPHRIES, R.N.** 1977a. The development of vegetation in limestone quarries. *Quarry manage. & Prod.*, 4, 43-47.

**HUMPHRIES, R.N.** 1977b. An ecological approach to the revegetation of limestone quarries in the United Kingdom. In: *Limestone quarries: responses to land use pressures*, edited by E.J. Perry and N.A. Richards, 2-39. Jamesville: Allied Chemical Corporation.

**HUMPHRIES, R.N.** 1979. Landscaping hard rock quarry faces. *Landscape Des.*, 127, 34-37.

**HUMPHRIES, R.N.** 1980. The development of wildlife interest in limestone quarries. *Reclamation Review*, 3, 197-208.

**KLEMOW, K.M. & RAYNAL, D.J.** Population ecology of *Melilotus alba* in a limestone quarry. *J. Ecol.*, 69, 33-44.

**PARK, D.G.** 1981. Seedling demography in quarry habitats. *This Symposium*

**SKALLER, M.P.** 1971. The ecology of natural plant colonisation in a limestone quarry. In: *Limestone quarries: responses to land use pressures*, edited by E.J. Perry and N.A. Richards, 40-71. Jamesville: Allied Chemical Corporation.

# THE RE-CREATION OF CONSERVATION VALUE IN MINERAL WORKINGS

**C G DOWN**
*Department of Mineral Resources Engineering, Imperial College, London*

## INTRODUCTION

The conflicts between mineral development and conservation interests, especially though not exclusively biological conservation, are well known in principle but scarcely solved in practice. When such conflicts arise, the "solution" is usually to permit the development and damage (or eliminate) the features of conservation value, or to refuse the development and preserve the competing values. Such a stark choice leaves very little room for any compromise. Indeed, the most that can normally be achieved is a slight alteration in either the location of some components of the development, or its timing, the latter in order to permit last-minute "rescue" studies to be performed. Archaeologists who carry out rescue digs in advance of new roads, or on building sites, will be familiar with this dilemma. One cannot re-site an office block at the last minute any more than a new orebody can be discovered to order in some less sensitive location.

The precise nature of the problem and the ways of handling it vary from country to country. At one extreme might be placed the North American implementation of a wilderness concept which debars mineral or other development. Such a course certainly avoids the conflict, though it may well create difficulties in other directions. At the other extreme might be placed countries such as the United Kingdom, Malaysia and some African nations. This apparently disparate grouping includes nations which have little if any land legally debarred to mineral operations, but fairly extensive areas of national parks and other designated lands within which there is some degree of presumption against mining and wherein the burden of proof of the need for the mineral usually lies squarely upon the developer.

Such situations are "large scale" in the sense that objections to mineral development are not usually based so much upon its local impact, however severe, as upon the principles involved and the "thin end of the wedge" fear. This paper considers a relatively local aspect of this type of conflict, namely the physical destruction of habitats deemed of biological conservation value by mineral (or other) development. Such conflicts arise frequently in the U.K. where an abundance of conservation sites (often only one or two hectares in extent) exists.

Although the judgement has already been stated in terms of stark choice between the existing habitat or the new land use, in fact it is becoming increasingly apparent that intermediate possibilities do exist. This is not to suggest that anything like a perfect compromise can as yet be found, but it nonetheless seems that the granting of permission for mining upon land of conservation value need no longer entail total loss of that value. At least four possiblities can be suggested:

1. To plan reclamation of the site such that natural recolonisation can take its course.
2. To accelerate and direct natural recolonisation of the site.
3. Deliberately and rapidly to re-create the valuable ecosystem by planting, either at the mine site or elsewhere.
4. Before development commences, to transplant the whole ecosystem to another place.

The first two possibilities are matters for other speakers, and this paper is devoted to the last two.

## RE-CREATION

Recolonisation is applicable largely to the reclamation phase of a quarry's life. In contrast, re-creation and transplanting are usually of more relevance to the pre-quarrying stage, since they are both likely to depend upon the preservation by re-location of vegetation that would otherwise be destroyed as a result of the quarry development.

As discussed by other speakers, it is possible to re-create upon quarried land a reasonable imitation of an herbaceous ecosystem (such as grassland) by selective planting coupled with natural recolonisation. Such an ecosystem could of course equally well be re-created upon undisturbed land, although there would seldom be any point in doing so. However to re-create rapidly a community such as a mature woodland is a task of a very different nature and magnitude. It has not, to the writer's knowledge ever been achieved (nor indeed have many attempts been made), but the writer has conducted work towards this end.

The problem can be defined as: how best to re-create or establish a mature woodland, quickly, on a new site, such that the woodland is a reasonable replica of the original. Three strands of thinking have been applied in parallel. First, a good model of the original was required. Accordingly a suitable woodland was studied by conventional ecological survey techniques, paying particular attention to soil characteristics, species composition, species associations, distributions and patterns, and light attenuation by the woodland canopy. The result of this survey was a good understanding of the wood as a whole, allowing selection of the major community types and patterns that should be imitated when re-planting.

The second consideration was to obtain rapid shading since shade is so clearly a major factor in woodland economy, and without it ground flora could not easily be perpetuated. Although artificial shading could be employed at the new site, it was decided eventually to try and achieve shading by the direct transplanting of mature trees, up to about 15m in height. Some were transplanted whole (using only an hydraulic backhoe for excavation, and conventional trucks for transport) without any prior preparation, while others were coppiced back to stumps to reduce their water demand at the reception site. Using nothing more than readily-available quarry plant, and operators without any previous practice, it proved possible to excavate, transport and replant such trees over a distance of 5km at a rate of about twenty per 10h day (Plate 8).

The third element was the ground flora. Collections were made of both common (typical) species, and rarities, in all cases as vegetative material and not as seeds, for propagation in a nursery.

Since 1981 is only the third growing season, speculation as to the results of these trials is premature. However, initial indications are that the trees have survived the very crude methods used, while propagation of the ground flora has been most successful. While there may well be other ways to achieve the objective, the approach adopted above has proved practical and workable to date.

## TRANSPLANTING

Although re-creation as described above involves transplanting, the operation is very different when applied to herbaceous communities such as grassland. These present the possibility of moving entire areas of ecosystem from one place to another. Such transplanting of turf is not of course new, having been long-established for purposes such as domestic lawns and sports fields; the challenge in the present context is to develop the method into an industrial-scale operation.

The approach adopted by the author has been governed by the same principles as the re-creation work described above: namely to employ only commonly-available mining machinery and operators not specially trained. Experiments set up since 1977 have been similar in essentials. The machine found most successful is a front-end loader, either rubber-tyred or crawler-mounted. The bucket (preferably smooth-lipped and not toothed) is used to cut out pieces of turf the size of the bucket floor (typically about 2.5 x 1.2m and up to 0.5m thick). The loader then transports each turf in succession to the designated reception site, and there deposits it (Plates 9-10). Turf sources for the trials have been recently-sown pasture, and moorland, with reception sites being on tips of coal waste and china clay waste respectively. Trials have been 5 x 3m and 5 x 12m at the largest plots used to date.

Although this procedure might appear slow, over distances of 1km it has been possible to achieve transfer rates equivalent to 1ha per 60-80 days. This is primarily limited by the relatively small bucket size available, and this rate could easily be increased to 1ha per 20-30 days using larger machinery. This would imply costs of around £3000/ha, or within the same range as the more expensive conventional seeding techniques, such as hydroseeding. Thus, although originally envisaged as a means of preserving threatened valuable plant communities, it may be that the technique could also find application in normal landscaping work.

This work has to date proved the technical and economic feasibility of large-scale transplanting. The most important question has yet to be answered: does the vegetation it is desired to preserve actually survive the transplanting process unchanged? Initial indications are favourable, but much remains to be ascertained and, indeed, understood.

## CONCLUSIONS

There seems little doubt that in the United Kingdom the preservation and enhancement of conservation interest at mineral workings will become of increased importance. This arises not only because of the continued loss of natural habitat, but because of the demonstrable value of abandoned mineral workings as sites of scientific interest. The approaches and techniques discussed above are neither wholly distinct from one another, nor mutually exclusive. They do however have in common the fact that important aspects of them are yet to be understood and much further work remains to be performed before any can be reliably employed at mineral operations.

## ACKNOWLEDGEMENTS

Many organisations and individuals have provided important help in formulating the ideas discussed, and in implementing field trials. I particularly wish to acknowledge my colleague Dr Alan Morton, with whom most of the practical work mentioned has been conducted.

*Plate 8 Transplanting a hornbeam tree* **Carpinus betulus** *into a prepared pit at the reception site.*

*Plate 9 Crawler-mounted front-end loader undercutting turf (***Calluna, Ulex, Agrostis, Molinia***) prior to transporting it to a reception site.*

*Plate 10 Example of completed trial area.*

# A RECLAMATION STRATEGY FOR QUARRYING

**N J COPPIN**
*Environmental Advisory Unit, Department of Botany, University of Liverpool**

Quarrying can have many effects on the environment so it is important to develop some environmental protection and reclamation programme within the life of the working. A reclamation strategy should allow for all the possible alternatives of after-uses and site conditions. The differences between operations are often very great, and each site must be considered on its own merits.

The basis of all landscape reclamation schemes will be made up of two parts, *engineering* and *biological* reclamation; the job of the landscaping specialist (whether Landscape Architect or Ecologist) will be to bring these two parts together. Engineering reclamation will be concerned with the creation of suitable landforms, and must account for stability of tips and rock faces, drainage, etc. Biological reclamation is concerned with establishing and maintaining vegetation on the new landform. Both these aspects must overcome problems with the materials available, and the operational requirements of the quarry. They must also take account of the needs of each other.

Vegetation will play a key role in providing screens, slope stabilisation and erosion control, soil improvement, visual improvement, and lastly but perhaps most important, providing some satisfactory after-use. Spoil, waste and overburden can also play a key role in treating pits and quarries. They are not always merely by-products that must be disposed of as cheaply and simply as possible. They can be a valuable resource with which to build a landscape: improving access and slopes in pit areas, providing backfill around faces, covering landfill, maintaining adequate drainage and constructing screening bunds around stark quarry landscapes. Some mineral extraction industries, of course, such as china clay, produce such large quantities of spoil that its disposal is a major problem.

The aim of a reclamation strategy should be to formulate a set of *landscape objectives* or *ecological goals*, and a flow diagram illustrating the steps involved is shown in Figure 13. These are discussed in more detail below.

1. The first stage in a new quarry development simply involves outlining the project: the geology and reserves, the size and scope of the workings, the production requirements and the land required for spoil disposal, etc.

2. Site survey and appraisal. This should identify topographical features, soils and land use classification, climate, surface and subsurface water movements, landscape quality and local vegetation. The local vegetation will be particularly important if any nature conservation interests are involved. Sampling of the overburdens and resources in advance of extraction can identify potentially difficult materials and spoils and those with some potential as a new soil. Stripping and spoil disposal can then be planned in advance to take this into account.

3. Growth trials and experimental studies can look at suitable plant species, soil amelioration and after-care as necessary.

4. After-use. Here we define the alternatives available. The after-use must be compatible with the potentials of the soil materials available, otherwise the treatment and after-care costs will be prohibitive. It must also match local land use needs.

*present address: *Robinson, Jones Partnership Ltd, 46 Clwyd Street, Ruthin, Clwyd.*

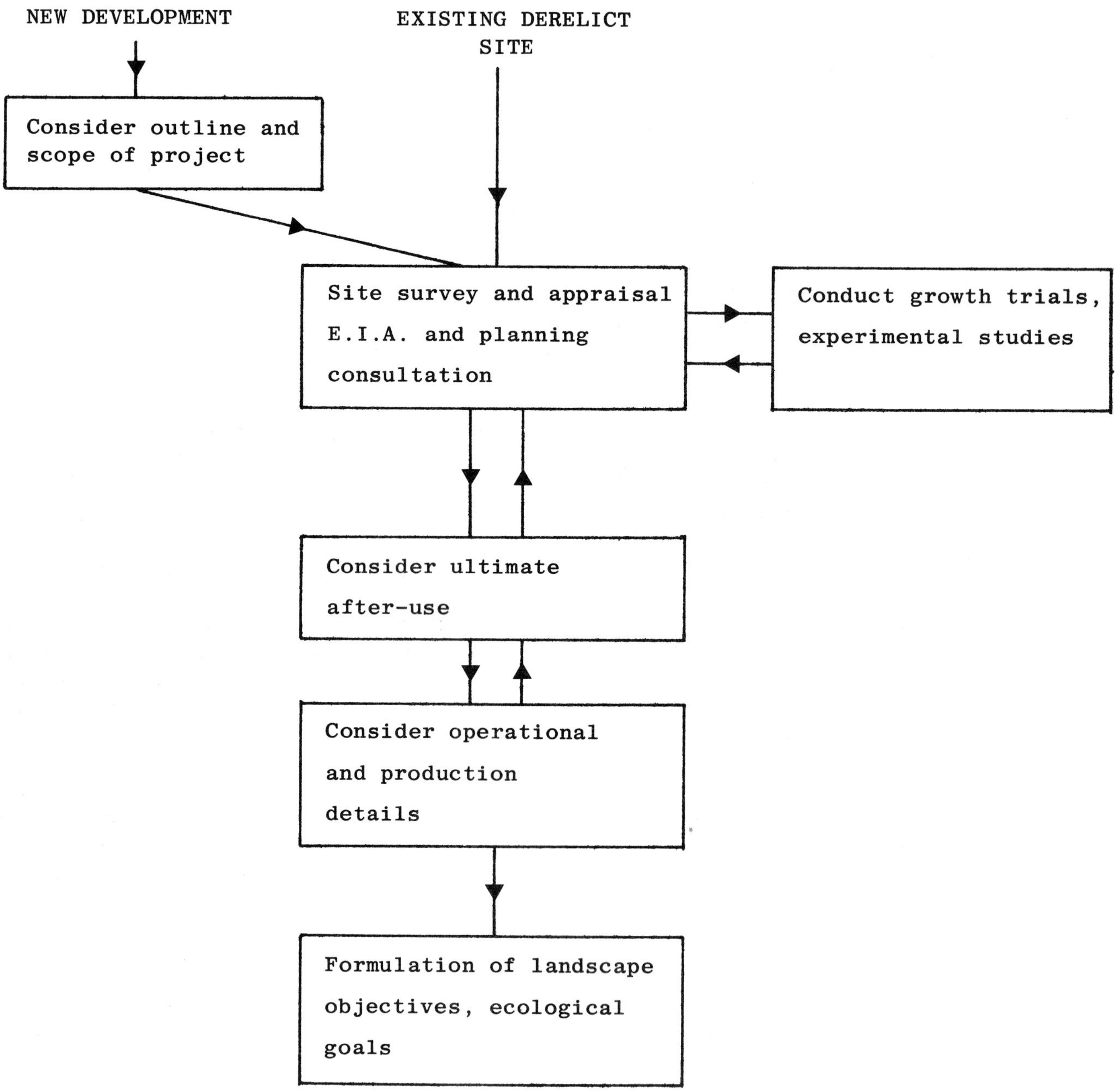


*Figure 13* *Stages in the formulation of landscape objectives for quarrying reclamation*

It is most important from the mineral operator's point of view that the eventual after-use of a new development is determined as early as possible. For long-term, hard rock quarries, though, it will only be possible to define a range of possibilities at the outset and decisions are taken as the workings progress. Local and regional land use priorities may change over time, so long term operation must remain fairly flexible in their final objectives.

The after-use possibilities usually associated with quarries are:

Housing and industry
Sport and intensive recreation
Agriculture and forestry
Non intensive recreation and country parks
Wildlife, conservation and eduction
Water storage and management
Landfill and waste disposal (temporary)

Some of these are illustrated in Plates 11-14. Very often two or more of these can be successfully developed for multiple or successional use, especially in large quarry sites where there is a wide variety of topographical features - floors, rock faces, benches and terraces, wet areas, etc. Shallow progressive workings with continual replacement of overburden or fill need only produce minimal disturbance. These can be fairly quickly returned to productive use, usually what was there originally.

Having considered all the previous points, we should now have a fairly clear idea of the landscape objectives and ecological goals. We will know what types of landforms we require, and the type of vegetation: its function productivity and management. We can then develop the working plans of the quarry or reclamation scheme.

This meeting has considered at great length the contribution that derelict quarry sites can make to the conservation of locally and nationally important plant and animal communities. It is clear that nature conservation, not just for rarities but wildlife generally, has great potential as an after-use. As yet, however, there are few practical guidelines as to how new workings can be developed in this way. This is in contrast to the great strides made recently in techniques for restoring quarries to agriculture, forestry and recreation. In order that nature conservation and wildlife should not be left out, the quarrying industry needs guidance from ecologists on the criteria for selecting suitable sites, and the techniques of site treatment to encourage wild species. There is a danger that, with modern planning and pressures to reclaim everything, a future source of wildlife refugia will completely disappear.

## ACKNOWLEDGEMENTS

This paper was written whilst the author was working on a project at the University of Liverpool, Environmental Advisory Unit, Department of Botany, sponsored by the Mineral Industry Research Organisation. It expresses the author's own personal views, which are not necessarily those of the Mineral Industry Research Organisation or the member companies.

*Plate 11 Restoration of a sand pit to agricultural use, using stockpiled topsoil. Part of the quarry face is still visible at the far side.*

*Plate 12 Use of a wet sand and gravel pit for recreation as a country park. (Reproduced by courtesy of Amey Roadstone Corporation Ltd)*

*Plate 13 An old brick-clay pit left as derelict. This has colonised naturally and is extensively used by local people for walking and motor cycle scrambling.*

*Plate 14 An old limestone quarry in the Peak District which is now a local nature reserve. Many rare limestone grassland plants have colonised the dumps of spoil on the quarry floor.*

# RESTORATION SCHEMES: AN INDUSTRIAL VIEWPOINT

**P E GAWN**

*Amey Roadstone Co. Limited, Lands Department, The Ridge, Chipping Sodbury, Bristo*

This paper aims to complement preceding ones by outlining the criteria considered and processes often followed within the quarrying industry when preparing schemes of working, restoration, and after-use. I am deliberately relating my comments in the main to those factors which have a bearing upon the re-creation of 'natural' habitats as opposed to restoration for uses such as agriculture, forestry, or other 'development'. In conclusion, I will say a few words concerning multiple after-use of quarries.

Background considerations include:

1. Geological and known physical constraints to the site. The former involves consideration of geological data obtained by means of drilling surveys and will include overburden ratios, projected face characteristics and stability, etc. Physical constraints mean restrictions imposed on the site by way of ownership limitations, neighbouring 'sensitive' development, hydrological considerations, etc.

2. Commercial opportunities for the land once worked, and thus the residual value of the site.

3. The visual and ecological significance of the site and its surroundings, obtained by established survey techniques.

4. Stated Planning criteria for the site by references to the Planning Authorities and known requirements of consultees such as MAFF, Water Authorities, and in the case of SSSIs the NCC.

5. The practical constraints and opportunities presented by the site ie likelihood or otherwise of the occurrence of unsaleable by-products; operational characteristics such as silt disposal; opportunities for the utilisation of available mobile Plant; the identified need for the creation of strategic screening bunds, tree planting, etc. and their subsequent management.

6. An assessment of the effects of the likely timescale involved in development and completion of the proposed scheme. The need for definition of objectives at an early stage, whilst allowing for flexibility of detail throughout the life of the working, is an important aspect under this heading.

7. The cost considerations of the scheme resulting from a synthesis of the above.

A scheme of working and restoration is prepared after considering the above together with all the ancillary day to day production and commercial aspects. Whenever possible, opportunities will be identified at this stage for building in a measure of progressive restoration, since industry fully recognises the general environmental benefits and economies to be realised by utilising Plant when it is available.

At this stage also, the scheme will often include broad guidelines as to the way in which the overall objectives will be achieved, ie creation of soiled benches, anticipated range of species to be utilised, subsequent management arrangements, etc. (Plate 15).

Largely due to the prolonged timescale involved in the development and restoration of a quarry, certain day to day problems may be anticipated to occur periodically. Typical of these are the following:

1. Possible changes in commercial practice and opportunity, eg development of new technical processes with their attendant by-products; local needs for a waste disposal site, etc.

2. Unforseen changes in the geological structure of the site.

3. Variations in marketing opportunity for various products (and thus the relative volumes of wastes and possibly unsaleable by-products).

4. "Political" changes, and a turnover in personalities amongst those concerned with the site and its restoration.

From this one can see that the factors which may influence the realisation of any restoration scheme are diverse and may not at first sight be directly related to the active restoration process. It is therefore necessary for anyone concerned with the working and restoration of quarries (both within the industry and from outside) to bear in mind the following:

1. The necessity to develop a full appreciation of the industry's problems, and to view the quarrying operation as a total entity, ie not to concentrate upon an immediate restoration detail in isolation from the numerous other factors involved.

2. The need to develop a positive approach to restoration problems and a readiness to take advantage of opportunities as and when they occur.

3. The need to develop as full an appreciation as possible of the commercial considerations involved in the development and running of a modern quarry (and a readiness to promote commercial opportunity should it arise).

4. The need to develop a flexible approach with regard to definition of restoration requirements, balanced by a knowledge of when to define strong principles which should not be negated.

The above may appear far removed and alien from the discussions of ecological principles and detail which has formed the basis of much preceding discussion during this Workshop. It is my submission however, that this is far from the case since I believe that valid arguments in favour of promoting habitat creation and conservation interests can only be developed on the basis of an appreciation of all the factors which contribute to the operation of a modern quarry. It is also my personal belief that on the majority of sites such interests can best be served by the formulation and execution of schemes which incorporate conservation principles alongside the development of other 'commercial' activities such as agriculture, industrial or leisure development. Owing to the various demands of society (and not intransigence on the part of quarry operators), opportunities for the reclamation of today's quarries for exclusive use as nature reserves will most likely be restricted to those sites which are remote, have poor access, and are small in scale. There may be exceptions, of course, for those which are demonstrated to possess unique geological features or habitats for endangered species.

Whilst I should emphasise that the views expressed above are essentially my own, I believe they reflect those of ARC and those of the industry at large.

*Plate 15 Soiling and planting on a remnant quarry bench in order to enhance visual amenity and habitat diversity*

# DISCUSSION

**CHAIRMAN: A D BRADSHAW**
*Department of Botany, University of Liverpool*

Perhaps the main reason why this workshop has been held is because with the passage of time, quarries develop considerable wild life value. But it needs to be pointed out that this value is entirely fortuitous, as the result of natural processes, and its development is very slow. It seems crucial therefore that we should have a positive restoration policy for native vegetation, not only because it will provide amenity and conservation but also because it can save on maintenance costs. This means that merely the construction of appropriate land forms, with perhaps inoculation of important species, may not be enough, even with the passage of time. Either wholesale removal and reinstatement of turf, or careful soil replacement with seeding may be successful if done with great care. Otherwise proper attention must be paid to ecosystem building processes, though in this case the end product will vary depending on the chemical and physical conditions of the site and the season. This can be seen as a problem if it is necessary to make sure that a particular species is maintained, but not if the aim is for general diversity. Major differences in rates of weathering and in potential for revegetation occur between calcareous and other rocks, and restoration schemes need to be appropriate to the locality.

Usually, no one treatment will be totally effective or appropriate to a whole site, and it may be better to consider multiple land use so that maximum benefit can be gained from the site. This means that it is essential that decisions on land use are made early, ideally when the planning application is made. The industry could welcome constructive suggestions at this stage, especially if they help to smooth the passage of the application for development by offering one, or more, end uses of potential value to the local population. Sometimes, however, it may not be possible to see all the potential of a quarry development before it is initiated, so ecologists and conservationists must be prepared to provide positive ideas at all stages.

It is commonly thought that ecologists are only interested in preservation and will either object to any disturbance or request the *re-creation* of communities which are threatened. Although these courses of action may be necessary for certain sites, it is obvious from the workshop that ecologists should make clear that they are also concerned about *creation* of new communities. Quarries are valuable in providing primitive, skeletal, habitats, which, because they represent early stages in succession, can be of interest in their own right, as well as providing habitats for species which will not stand competition.

Early stages of a succession will lead eventually to later stages, sometimes very rapidly. As a result the ecological value of the site may be lost unless it is properly managed. Proper management is also necessary to ensure successful ecosystem development in the first instance. The management needs of quarry sites being used for wild life must, therefore, be understood and taken into account from the start, whether this means fertiliser application, sheep grazing or some other operation. If management is not going to be possible, its consequences must be accepted, and a different end point chosen from the outset.

Before we can harness the ecological potential of quarries, there are many specific ecological problems that still need to be tackled. If we really are to understand how succession in quarry situations will progress, or how we can retain particular species, we need to know a great deal more about the factors which determine dispersal, establishment and subsequent growth.

We need to know more about soil/plant relationships in selected soils and the development of properly functioning soils. There is also too little known about the role of animals in ecosystem development. Quarries also raise the problem of minimum areas for conservation of species.

Man-made habitats such as quarries have not, until recently, really been ecologically respectable. Yet they offer remarkable opportunities to study fundamental ecological attributes of organisms and to understand the processes which contribute to the development of ecosystems. At the same time, they are of proven conservation value. This workshop makes it clear we are beginning to recognise their academic and practical ecological potential.

## ADDRESSES OF CONTRIBUTORS AND PARTICIPANTS


Mrs Penny Anderson, Gwynfa, Buxton Road, Chinley, Stockport, Cheshire SK12 6DR

Dr A J M Baker, Department of Botany, University of Sheffield, Sheffield S10 2TN

Dr N G Bayfield, ITE, Hill of Brathens, Glassel, Banchory, Kincardineshire AB3 4BY

Prof A D Bradshaw, Department of Botany, University of Liverpool, PO Box 147, Liverpool L69 3BX

Mr N J Coppin, Robinson, Jones Partnership Ltd, Ty Gwyddon, 46 Clwyd Street, Ruthin, Clwyd LL15 1HW

Dr B N K Davis, ITE, Monks Wood Experimental Station, Abbots Ripton, Huntingdon PE17 2LS

Dr J P Doody, NCC, Godwin House, George Street, Huntingdon PE18 6BU

Dr C G Down, Department of Mineral Resources Engineering, Imperial College of Science & Technology, London SW7 3BP

Mr J Essex, Thurrock Borough Council, Council Offices, Grays, Essex RM17 6SL

Mr B G Finegan, Department of Applied Biology, University of Cambridge, Pembroke Street, Cambridge CB2 3DX

Mr P E Gawn, Amey Roadstone Co Ltd, Lands Department, The Ridge, Chipping Sodbury, Bristol BS17 6AY

Dr J E G Good, ITE, Bangor Research Station, Penrhos Road, Bangor, Gwynedd LL57 2LQ

Miss Hilary J Gray, Department of Botany, University of Liverpool, PO Box 147, Liverpool L69 3BX

Dr P J Grubb, Botany School, University of Cambridge, Downing Street, Cambridge CB2 3DA

Dr H J Harvey, Department of Applied Biology, University of Cambridge, Pembroke Street, Cambridge CB2 3DX

Mr D W H Heywood, Abax Hemerdon Ltd, Hemerdon Mine, Plympton, Plymouth, Devon

Mr J Hodgson, NERC Unit of Comparative Plant Ecology, Department of Botany, Sheffield S10 2TN

Dr M D Hooper, ITE, Monks Wood Experimental Station, Abbots Ripton, Huntingdon PE17 2LS

Dr R N Humphries, Department of Applied Biology, University of Cambridge, Pembroke Street, Cambridge CB2 3DX

Dr M S Johnson, Department of Botany, University of Liverpool, PO Box 147, Liverpool LE69 3BX

Dr R H Marrs, ITE, Monks Wood Experimental Station, Abbots Ripton, Huntingdon PE17 2LS

Dr T McNeilly, Department of Botany, University of Liverpool, PO Box 147, Liverpool LE69 3BX

Mr D R Morrey, 31 Westfields, Carr Lane, Spennymoor, Co Durham

Mr D G Park, ITE, Monks Wood Experimental Station, Abbots Ripton, Huntingdon PE17 2LS

Mr C E Ranson, NCC, Centurion House, St John's Street, Colchester, Essex CO2 7AH

Miss Sally Richardson, Department of Applied Biology, University of Cambridge, Pembroke Street, Cambridge CB2 3DX

Dr R D Roberts, Department of Biology, University of Essex, Wivenhoe Park, Colchester, Essex CO4 3SQ

Miss Helen V Smith, Department of Applied Biology, University of Cambridge, Cambridge CB2 3DX

Mr D Wells, NCC, Godwin House, George Street, Huntingdon PE18 6BU

Mr T C E Wells, ITE, Monks Wood Experimental Station, Abbots Ripton, Huntingdon PE17 2LS